Topics in Applied Physics Volume 21

Topics in Applied Physics Founded by Helmut K.V.Lotsch

Solid Electrolytes

Edited by S. Geller

With Contributions by
S. Geller   L. Heyne   J. H. Kennedy   B. B. Owens
J. E. Oxley   A. F. Sammells   H. Sato   W. L. Worrell

With 85 Figures

Springer-Verlag Berlin Heidelberg New York 1977

Professor Dr. *Seymour Geller*

University of Colorado, Department of Electrical Engineering,
Boulder, CO 80309, USA

ISBN 3-540-08338-3 Springer-Verlag Berlin Heidelberg New York
ISBN 0-387-08338-3 Springer-Verlag New York Heidelberg Berlin

Library of Congress Cataloging in Publication Data. Main entry under title: Solid electrolytes. (Topics in applied physics; v. 21). Includes bibliographies and index. 1. Electrolytes. 2. Solids. I. Geller, Seymour, 1921-. QD565.S566. 541′.37. 77-21873

Monophoto typesetting, offset printing and bookbinding: Brühlsche Universitätsdruckerei, Lahn-Gießen
2153/3130-543210

Preface

Solid electrolytes have been known for many years, going back to Faraday at least. Solid electrolytes are materials that have electrolytic conductivity in the solid state. There are several types of these materials as will be seen in the various chapters of this book. These materials have received the attention of researchers for some time, but not nearly the emphasis that they have been receiving in the last decade and particularly in the last two or three years. This is at least partially a result of the increasing interest in new energy sources.

Our purpose in writing this book is to present a concise view of some of the salient topics in this field. It is hoped that the book will give the researchers in the field and others interested in it some good ideas about these specialized topics and that they find the book useful in their own work. It should help also to clear away some misconceptions which tend to enter a fast-growing field, and, it is hoped, discourage new ones.

Fortunately, the field at present has a youthful vigor and there appears to be considerable research yet to be done in it. We hope that this book will at least help to point to some of the paths that could be taken.

Following a brief introductory chapter, there is a chapter by H. Sato on theoretical aspects of solid electrolytes, in particular, those in which the conductivity is high and the enthalpy of activation of motion of the charge carrying ions is low. Theory might be said to be in its early stages at this time, but it will be seen that progress has been made.

Following this chapter there are two on the halogenide solid electrolytes, the first by S. Geller which emphasizes the crystal structural aspects of the origin of the electrolytic conductivity in these materials and the second by B. B. Owens, J. E. Oxley and A. F. Sammells which deals with applications of these materials.

The next chapter by J. H. Kennedy is on the specialized subject of the β-alumina solid electrolytes on which it is probable that most work is presently being done. This chapter is an overview of this work and includes sections on a large variety of scientific studies as well as on applications. The chapter by W. L. Worrell on oxide solid electrolytes in which the oxygen ions is the main charge carrier follows. These all have the fluorite crystal structure and are solid electrolytes at very high temperatures only. Again, the chapter ranges over scientific and applied aspects of these materials.

Finally there is the chapter by L. Heyne on the electrochemistry of mixed ionic-electronic conductors. All solids have some electronic (or hole) conductivity in a wide range. This chapter combines thermodynamic defect chemical and semiconductor physical principles into a solid state electrochemistry. The chapter is also largely theoretical, overlapping somewhat with other chapters. It deals largely with dilute point defect materials but includes those with large numbers of ionic carriers as well.

Boulder, Colo.   *S. Geller*
June, 1977

Contents

Contributors

Geller, Seymour
  University of Colorado, Dept. of Electrical Engineering,
  Boulder, CO 80309, USA

Heyne, Leopold
  Philips Research Laboratories, Eindhoven, The Netherlands

Kennedy, John H.
  Dept. of Chemistry, University of California, Santa Barbara, CA 93106, USA

Owens, Boone B.
  Medtronic Inc., 3055 Old Highway Eight, Minneapolis, MN 55418, USA

Oxley, James E.
  Gould Laboratories, Rolling Meadows, IL 60008, USA

Sammells, Anthony F.
  Rockwell International, Atomics International Division,
  Canoga Park, CA 91304, USA

Sato, Hiroshi
  Purdue University, School of Materials Engineering,
  West Lafayette, IN 47907, USA

Worrell, Wayne L.
  Dept. of Metallurgy and Materials Science, University of Pennsylvania,
  Philadelphia, PA 19174, USA

1. Introduction

S. Geller

There are two types of conductivity: electronic and electrolytic. The first does not involve material transport while the second does. Both occur in all states of matter to a greater or lesser degree. This book is concerned mainly with electrolytic conductivity in solids, but one cannot escape consideration of electronic conductivity, because no matter how low it may be, it is there, so to speak. This is not the case for electrolytic conductivity—a crystal must contain ions to have it, if it is there at all.

The literature contains many papers on solid electrolytes and there is a question concerning their classification. (In fact, there are some slight differences of opinion even among some of the authors of this book.) Many investigators have referred to dilute point defect ionic solids as solid electrolytes. Indeed, these are included as solid electrolytes in certain sections of this book. They occur mainly as a result of an equilibrium concentration of defects that exists in almost all crystals. The origin of the conductivity in such crystals is associated with the annihilation and recreation of such defects. Included among crystals of this type are the alkali halides, the silver halides (except for α-AgI), and many others.

My own point of view is that the aforementioned crystals are solid electrolytes "by accident". Generally they have very low conductivities and high carrier activation enthalpies. True solid electrolytes are different. This is clearly indicated in various parts of this book which show the important association of the electrolytic conductivity phenomenon in ionic solids with their crystal structures. In these structures, it can be shown in every case that there are pathways for the carriers that are built into the crystal structure. Among the structural characteristics (all of which are discussed in later chapters) is the excess number kf available sites for the carriers relative to the number of carriers, and usually it is clear from the structure that the network of the pathways results from face-sharing of polyhedra formed by anions in the case of cation conductors, and both anions and cations for cases in which the charge carrier is the anion.

There are certain terms that have been used in discussing true solid electrolytes that are somewhat misleading. Examples are "molten" or "liquid" sublattice referring to the condition of the mobile charge carriers. These terms might lead to the belief that these charge carriers are distributed at random in all sorts of interstices in the crystal structure of the solid electrolyte. This is not the case at all as especially emphasized in Chapter 3 (see especially Sect. 3.1.3)). The above nomenclature may be found in some places in this book. However, it should be kept in mind that it is only a shorthand way of describing the situation and that such terms are used rather loosely.

Another such expression is "superionic conductor". The use of this term has become rather widespread; nevertheless it is not used in this book because it is grossly misleading, so much so that it has even appeared in a reversed form as "ionic superconductor" (e.g., [1.1, 2]). Needless to say, there is not the remotest relation of these materials to superconductors. There is no need or justification for the term; what is *called* the "superionic conductor" *is* the *true* solid electrolyte. There is no scientific justification for a "superion" entity nor liguistic justification for an earlier used variant "super ionic conductor", which was apparently meant to be interpreted as an ionic conductor which is "super" (not literally a word). Because of the importance of the development of the electrochemistry of the mixed ionic-electronic conductor from the point defect approach, the dilute point defect conductor is included as a solid electrolyte in Chapter 7. In all cases, the word "true" will be omitted even when referring to true solid electrolytes. This should actually not result in any ambiguity, because the reader will readily discern the correct situation.

There is another, but useful, term, namely "fast ion transport" as in the title of the book edited by Van Gool [1.3]. This refers to the relatively high ionic mobility associated with the ionic carriers in the true solid electrolytes. (Not all the papers in Van Gool's book are actually in this category.)

While throughout the book, real materials are considered even in the more theoretical chapters, i.e., Chapters 2 and 7, the emphasis on materials is greatest in Chapters 3–6 inclusive. The ionic conductivity of crystalline materials is strongly related to their crystal structures; consequently there is much reference to these throughout Chapters 3, 5, and 6.

The choice of topics for this book was dictated essentially by the literature, i.e., as an indication of the interest which they create, and the limitation of book size. The greatest interest appears to be in the β-aluminas, and next the halogenides (probably more on the AgI-based solid electrolytes than any other group) and finally the fluorite-type oxide electrolytes. There are, of course, others such as those reported by Goodenough et al. [1.4] that do not fit the categories listed. In the categories listed, there have already been applications, some, as in the case of the oxide solid electrolytes, for a rather long time. Chapter 4 is devoted to applications of halogenide solid electrolytes and sections of Chapters 5 and 6 discuss applications.

Tables and figures are included mainly to make specific points. There are now several papers in the literature (e.g., [1.3; 2.3; 3.3, 69, 107]) that give lists of electrolytes and their specific properties and these are not repeated here.

We expect (and hope) that this book will be useful to all interested in solid electrolytes or related fields. It is possible that in reviewing some aspects of the solid electrolyte field an important paper has inadvertently been overlooked. More important, it is impossible to give, in a review, all the flavor of an original author's paper. In fact, it could happen *in any review* that the original author's view is even unintentionally misrepresented. In any case, we encourage the reader who intends to go more deeply into the field to read original papers, especially if some doubt occurs concerning any of the points discussed.

References

1.1 A.Guinier: Phys. Today, **28**, 23 (1975)
1.2 J.M.Reau, C.Lucat, G.Campet, J.Claverie, J.Portier, P.Hagenmuller: New Fluor Ion Conductors. In *Superionic Conductors*, ed. by G.D.Mahan, W.L.Roth (Plenum Press, New York, London 1976)
1.3 W.VanGool: *Fast Ion Transport in Solids* (North Holland, Amsterdam-London 1973)
1.4 J.B.Goodenough, H.Y.-P.Hong, J.A.Kafalas: Mat. Res. Bull. **11**, 203 (1976)

2. Some Theoretical Aspects of Solid Electrolytes

H. Sato

With 18 Figures

In this chapter, we discuss some theoretical efforts to explain two basic features of solid electrolytes: the origin of the formation of the substantial disordering required for the electrolytic conductivity in solid electrolytes and the transport mechanism in such disordered structures. Greatest emphasis is given to the explanation of a theory which deals with essential problems of ion transport.

2.1 Sublattice Disorder

In nearly perfect crystals, atomic (ionic) diffusion is always connected to the existence of lattice defects [2.1–3]. The most common types of diffusion in crystals are diffusion through vacancies (vacancy diffusion) and diffusion through interstitial sites (interstitial diffusion). In ionic crystals, such motion of ions under the influence of an external field creates a net ionic current. The ionic conductivity σ is generally described by an Arrhenius equation

$$\sigma = (C/kT)\exp(-u/kT) \tag{2.1}$$

where u is the activation energy of ion motion, C the pre-(exponential) factor, k the Boltzmann constant, and T the temperature. C is given by

$$C = (1/3)(Ze)^2 n d^2 w_0 \tag{2.2}$$

where Ze is the charge of the conducting ion, n is the density of defects (the density of vacancies in vacancy diffusion or the density of interstitial ions in interstitial diffusion), d is the unit jump distance of the ion, usually the closest ionic pair distance, and w_0 is the attempt frequency. The corresponding diffusion coefficient D is defined by

$$D = D_0 \exp(-u/kT), \tag{2.3}$$

$$D_0 = C/(Z^2 e^2 n) \tag{2.4}$$

and hence

$$\sigma = n(Ze)^2 D/(kT). \tag{2.5}$$

The last equation is called the Nernst-Einstein relation. The derivation of these relations can readily be made by the application of the random walk theory [2.1–3] based on the assumption that n is small.

Let us look specifically at vacancy diffusion. It is immediately clear from (2.2) that, to have a large ionic conductivity, the number of vacancies must be large enough so that the effective number of ions contributing to the diffusion can be large. In other words, if one can find crystals in which the number of available sites is greater than the number of the diffusing ions so that the ions can distribute over these available sites at relatively low temperatures, one can expect a high ionic conductivity. For such solids to be rigid, the disorder should occur only in one of the sublattices where diffusing ions exist.

Let us examine the situation with typically known solid electrolytes. One of the best known examples is α-AgI. Here, the iodide ions form a rigid, body-centered cubic sublattice while the small Ag^+ ions (two per unit cell) distribute homogeneously in the twelve tetrahedral interstitial sites of the iodide lattice. (For details, see Sec. 3.1.1.) Based on x-ray diffraction results with polycrystalline specimens, *Strock* [2.4] first pointed out that Ag^+ ions were distributed statistically over a large number of interstitial positions, and this situation was then thought to be largely responsible for the high ionic conductivity of α-AgI. β-alumina has similar characteristics with Na^+ the disordered cations. Na^+ ions exist in crystallographically loose layers which periodically modulate spinel-like layers of Al_2O_3 [2.5, 6]. However, the number of Na^+ ions in the layer is more than the number of equivalent crystallographic sites specified for Na^+ ions of an established crystallographic structure $Na_2O \cdot 11Al_2O_3$ (the Beevers-Ross structure [2.5, 6]), a feature often cited as "nonstoichiometry". Therefore, Na^+ ions are forced to distribute over a large number of available sites (not equivalent) which are more numerous than the number of Na^+ ions [2.6, 7]. Let us call such a situation "sublattice disorder". The words "cation disorder" [2.8], "liquid sublattice" [2.9], etc., also have been used with similar connotations.

In the more complex solid electrolytes, the conducting ions are distributed nonuniformly over crystallographically nonequivalent sites (see Chap. 3). However, the average activation energy of ionic motion is small relative to that of ordinary solids. In some "fast ionic conductors" (see Chap. 1), the activation energies are of the order of 0.1 to 0.2 eV. The smallness of the activation energy (and, at the same time, the smallness of the preexponential factor) is indeed a better criterion for characterization of fast ionic conductors than the ionic conductivity itself as will be made clear later. In some cases, the ionic conductivity changes very little [2.10], or even decreases [2.11] upon melting with similar activation energy and preexponential factor. This fact cannot really suggest that the sublattice disorder in the solid is structurally equivalent to that in the liquid. Nevertheless, this has been the origin of the term "liquid sublattice". Based on these experimental evidences, however, it is at least evident that the "sublattice disorder" is responsible for the characteristics of fast ionic conduction.

O'Keeffe [2.10, 12], on the other hand, empirically classified some ionic solids into three categories with respect to fast ionic conduction as shown in

Table 2.1. Examples of materials exhibiting the various types of transition (*O'Keeffe* [2.12])

Class I	Normal melting
	alkali halides, $PbCl_2$, $MgCl_2$, $CaBr_2$, YCl_3, etc. [2.10]
Class II	First order transition
	Cation conductors:
	Ag^+, Cu^+ salts e.g. CuBr, CuI, AgI, Ag_2S [2.8, 14]
	Anion conductors:
	LuF_3, YF_3, $BaCl_2$, $SrBr_2$, Bi_2O_3 [2.10]
Class III	Faraday transition (diffuse)
	Cation conductors:
	Na_2S [2.16], Li_4SiO_4 [2.17]
	Anion conductors:
	CaF_2 [2.10], SrF_2 [2.10], $SrCl_2$ [2.18], PbF_2 [2.19]
	LaF_3 [2.10], CeF_3 [2.20]

Table 2.1. A large number of ionic solids are found to undergo a transformation from a poorly conducting state at low temperatures to a highly conducting state with the activation energy of ionic motion and preexponential factor apparently characteristics of liquids at high temperatures. Class I solids, or those which might be called normal salts (typified by alkali halides), do not undergo such "nonconducting-conducting transformation" but become highly conducting only upon melting. The conductivity change upon melting is typically from $\sim 10^{-3}\Omega^{-1}cm^{-1}$ to $\sim 1\Omega^{-1}cm^{-1}$ and the entropy change of fusion is remarkably constant at about $10 \sim 12\,JK^{-1}/g\cdot atom$. On the other hand, some solid electrolytes, notably Ag^+ and Cu^+ salts, exhibit a first order phase transition to the high-conducting state. In particular, the entropy change in connection with this solid state transformation in these substances is comparable to the entropy of fusion, and the ionic conductivities both just below and just above the transition temperature are in the same range as for normal salts (Class I solids) just below and above the melting temperature [2.10]. These solids are classified as Class II solids. There is also a number of salts, in which the nonconducting-conducting transformation is not a first order transition as in Class II solids but is spread out over a substantial temperature range. Such salts are classified as Class III. Through the temperature range of the transition, the conductivity passes smoothly from values typical of normal salts to values typical of ionic melts as shown in Fig. 2.1. (See also Sect. 3.3.) The heat content H is likewise a continuous function of temperature over the whole range of the transformation, but the total heat content of the transformation is again consistent with the heat of fusion. The heat capacity $C_p=(\partial H/\partial T)_p$ shows a peak similar to that of a second order transformation although there are no data as yet to indicate any singularity in the temperature dependence of C_p. The transformation was named the Faraday transformation by *O'Keeffe* [2.12] because Faraday was the first to point out that some ionic solids existed which showed the nonconductor-conductor transition. The apparent similarity of the charac-

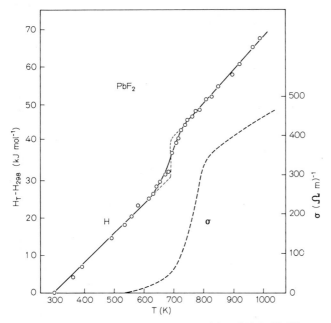

Fig. 2.1. The heat content and ionic conductivity of PbF$_2$ [2.13]

teristics of the nonconducting-conducting transition to those of fusion in both Class II and Class III solids has thus been emphasized [2.12].

If the characteristics of fast ionic conduction are due to the sublattice disorder, the existence of a transformation to a conducting state with temperature means that some structural changes or excitations occur with temperature which lead to a phase change. A possible structural change for the disorder which, at the same time, increases the conductivity is an excitation or a repopulation of conducting ions from their regular sites to other available interstitial sites. A detailed treatment of this process was given by *Sato* and *Kikuchi* [2.21] for the two dimensional honeycomb lattice with two equivalent sublattices A and B (Fig. 2.2). Here, A sites are the regular sites and the activation energy w is required to promote an ion from an A site to a B site. Such an excitation process (Frenkel defect formation) is, like Schottky defect formation, a typical "Schottky process" and the specific heat vs temperature curve gives rise to a Schottky-type broad maximum. Interactions among ions change the situation. If the nearest neighbor interaction (interaction between ions on neighboring A and B sites, respectively) is denoted by ε, the energy of the excited state E, based on the Bragg-Williams approximation, is proportional to $(w\varrho - \varepsilon\varrho^2)$, where ϱ is the fractional number of ions excited $(0 < \varrho < 1)$. If $\varepsilon < 0$, or a repulsive interaction (like the Coulomb interaction) exists, the occupation of the A sites is favored at low temperatures. The existence of repulsive interactions raises the temperature of the specific heat maximum and, at the same time,

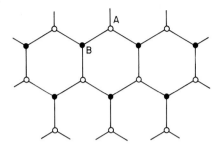

Fig. 2.2. Two sublattices A and B of two dimensional honeycomb lattice. A corresponds to the Beevers-Ross site and B to the anti-Beevers-Ross site in β-alumina

makes it sharper. In the limit of $|w| \ll |\varepsilon|$, on the other hand, the excitation becomes a second order phase change [2.21]. The calculation was made based on the pair approximation of the cluster variation method [2.22, 23] which is equivalent to the Bethe approximation.

If ε is positive, the excitation tends to reduce the energy of the further excitation, and the formation of the Frenkel defects becomes cooperative. Further, if such an interaction is strong enough, the excitation becomes eventually the first order transition. *Huberman* [2.24] introduced a large enough positive ε to derive such a first order transition to account for the transitions in Class II solids. He rationalizes the interaction as the effective interaction between an interstitial cation and a lattice vacancy it has left behind. It seems, however, difficult to justify the existence of such a strong attractive interaction between the Frenkel pair to promote the spontaneous formation of the defect, although, in his equation, an additional lattice loosening term as a result of excitation is included and this term apparently plays an important role. *Rice* et al. [2.25], on the other hand, interpret ε as the lattice loosening effect due to the interaction of interstitial defects with the strain field they induce. By minimizing the free energy $F(\varrho, T)$ of the form

$$F(\varrho, T) = \varrho(\varepsilon_0 - 1/2\lambda\varrho) - kT[\varrho \ln \varrho - (1 - \varrho)\ln(1 - \varrho)] \qquad (2.6)$$

with respect to ϱ, the equilibrium value of ϱ is obtained (Fig. 2.3). In this equation, ε_0 corresponds to w indicated above, whereas λ indicates the strength of the strain energy, and the second term of the equation is the entropy of mixing. Beyond a certain value of λ, a first order phase transition is predicted. For λ/ε_0 close to the critical value of 0.98, the transition to the disordered state is sharp enough to resemble the behavior of Class III solids. Whether or not λ can be so large as to induce the first order transition is yet to be determined.

The first order transition from the nonconducting state to the conducting state is generally accompanied by a lattice transformation. In AgI, iodide ions form close packed structures [either h cp (β-AgI) or fcc (γ-AgI)] at low temperatures while at the conducting state above 146 °C (α-AgI) have a bcc arrangement. Although the excitation of ions to interstitial sites depends on the crystal structure and a strong coupling between the Frenkel pair formation and the crystal structure is conceivable, the question whether or not the first order

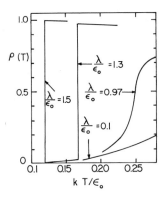

Fig. 2.3. The interstitial defect population as a function of temperature for increasing values of the interaction parameter λ. A first order transition is predicted for $\lambda/\varepsilon_0 > 0.98$ [2.25]

phase transition in Frenkel pair formation is directly connected to the lattice transformation requires further elucidation.

Many salts which belong to Class III solids have the fluorite or the CaF_2 structure (CaF_2, BaF_2, SrF_2, $SrCl_2$). It is known that, in the CaF_2 structure, the excitation of anions to interstitial sites occurs readily. In the CaF_2 structure, cations form a fcc lattice and anions are at the tetrahedral sites of the cation lattice. At high temperatures, anions are excited into the cubic sites, creating vacancies at the tetrahedral sites. The statistical treatment of the excitation can be made easily based on a similar scheme [2.26]. In view of the above treatment, it seems reasonable to expect that the transition in Class III solids is caused by such a disordering of ions and corresponds to the case for which γ in (2.6) is close to its critical value. *O'Keeffe* [2.12], on the other hand, suggests that this change is caused by a transition from "bound" states with vibrational modes to "free" states with translational modes of ions in view of quasifree motion of ions in the conducting state of fast ionic conductors (Sect. 2.2). A crude calculation by him [2.12] shows, however, that such a transition should have a wider temperature range than the actual experiments indicate.

In $RbAg_4I_5$, a break in the $\log \sigma T$ vs $1/T$ plot similar to those in Class III conductors has been observed at 209 K [2.27]. Measurements on specific heat and other physical properties of $RbAg_4I_5$ indicate that this change is, unlike that of Class III solids, a second order transformation [2.28, 29]. (See also Sect. 3.1.1.) A first order transition just like that of Class II solids is observed at $\sim$ 122 K in this material [2.30, 31], which suggests that the phase transition at 209 K is a phase transition in the conducting phase. As shown earlier, this type of change in the conducting phase can be explained as due to the order-disorder transformation among conducting ions and vacancies ($w = 0$ with a negative ε). The ordering creates preferential sites among otherwise equivalent sites, and the ionic conductivity decreases [2.32]. Based on his treatment of a lattice gas model of fast ionic conductors, *Mahan* [2.33] actually suggested that the anomaly in $RbAg_4I_5$ at 209 K was an order-disorder transition of conducting ions. Experimentally, the transformation at 209 K creates a minor change in the structure [2.34], but whether or not this change can be attributed to the theoretically discussed order-disorder transformation of conducting ions is not immediately clear. (Note that *Geller* [2.34] has shown that the transformation is structurally a disorder-disorder transformation.)

In addition to the disordering of ions which are constituent units of crystals, a fast ionic conduction can also occur in the form of interstitial diffusion. TiS_2, for example, has a layered structure consisting of Ti planes and S planes, and small alkali ions like Li can be intercalated between two layers of sulphur ions [2.35]. Li sites in the intercalated layer are equivalent to those of Ti in the Ti layer of $Li_x TiS_2$ and the concentration of intercalated Li ions x can be varied from 0 to 1 without disturbing the matrix too much (although $\sim 10\%$ expansion in the c-direction takes place as $x \rightarrow 1$) [2.36]. The diffusion of Li in TiS_2 is thus two dimensional. The intercalated plane, therefore, corresponds to an idealized model of fast ionic conductors in which conducting ions occupy the lattice points only partially. Similar compounds are also found in oxides [2.37, 38]. Some hydrogen conductors like Nb, Ta, V, etc., are three dimensional counterparts of TiS_2.

It has been shown that fast ionic conductors have some kind of sublattice disorder. Low activation energies of these substances are further attributed to their specific structural characteristics. Extremely low activation energies of some typical fast ionic conductors are, however, still hard to explain by structural features alone. It is an important theoretical task, therefore, to establish whether the magnitudes of the activation energies are solely determined by structural considerations (like that of one ion motion utilized in the random walk theory) or whether they depend mainly on the mechanism of ion transport and hence on the specific character of conducting ions. It is the purpose of later sections of this chapter to show that at least a part of the reduction in the magnitude of activation energy over that of the single ion motion can be attributed to cooperative motions of ions which occur in disordered systems.

2.2 Ionic Motion in Fast Ionic Conductors

In almost perfect crystals, the diffusion of ions takes place in the form of hopping. That is, ions spend most of their time in their respective potential wells and the dwell time of ions in the potential wells is long compared to the hopping time of ions to neighboring sites [2.39]. The ions are located mostly in their specific lattice sites and the attempt frequency ω_0 should be the ordinary vibrational frequency of crystal lattices or $\sim 10^{13} s^{-1}$. In fast ionic conductors, conducting ions are distributed over a large number of available sites (sublattice disorder) and the activation energy of ionic motion is low. This indicates that the potential wells in which the conducting ions are located are shallow. This situation also makes the attempt frequency low. At the same time, the amplitude of thermal motion becomes large as indicated by the results of diffraction experiments. Therefore, the motion of ions in these substances cannot, in general, be represented by a simple hopping diffusion.

For example, in α-AgI, recent diffraction results [2.40–42] indicate that the silver ions are preferentially found in elongated ellipsoidal regions of space centered at the tetrahedral interstices, extending in the $\langle 100 \rangle$ directions towards the neighboring octahedral sites. The Debye-Waller factor derived from such measurements corresponds to r.m.s. deviations of silver ions from their mean locations of more than 0.4 Å at 250 °C.

If ions are making such large-amplitude motion around many possible interstitial positions and, at the same time, translational motions from one site to the other, it should be possible to observe such motions as the high frequency response of the conductivity. In fact, unusually strong far infrared absorption [2.43] as well as intense Raman [2.44] and neutron scattering [2.45, 46] are observed in the frequency range less than 5–50 cm^{-1} of this substance, substantiating the existence of a slow, large-amplitude motion. Especially *Funke* et al. [2.45–47] analyzed their quasielastic neutron scattering data and came to a conclusion that silver ions in α-AgI might be described as a superposition of a translational jump diffusion and a large-amplitude local random motion. They analyzed the data in terms of dwell time τ_0 and the mean hopping time τ_1 assuming that the jump distance d is of the order of the lattice constant. This gave the value for $\tau_0 = 7$ ps and $\tau_1 = 15$ ps. In addition to α-AgI, measurements of Raman spectra for many fast ionic conductors have been carried out. These measurements all confirmed the existence of a liquid-like, slow large-amplitude motion in these substances.

The ratio of τ_1 and τ_0, or the ratio of the degree of diffusive motion with respect to the oscillatory motion, should be different from substance to substance, because this should depend on structure, especially on whether there are well defined stable ionic sites. An elementary discussion of the distribution of the above two degrees of freedom is conveniently carried out in terms of the velocity correlation function

$$z(\tau) = \langle v(t) \cdot v(t+\tau) \rangle \qquad (2.7)$$

and its Fourier transform

$$\tilde{z}(\omega) = 1/(2\pi) \int_{-\infty}^{\infty} z(\tau) \exp(-i\omega\tau) d\tau \qquad (2.8)$$

as is commonly done in the treatment of liquids [2.48]. The latter quantity is related to the diffusion coefficient D by

$$\tilde{z}(0) = \frac{1}{\pi} \int_0^{\infty} z(\tau) d\tau = D/\pi. \qquad (2.9)$$

Also, the integral over ω gives simply

$$\int_{-\infty}^{\infty} \tilde{z}(\omega) d\omega = \int_{-\infty}^{\infty} z(\tau) \delta(\tau) d\tau = \overline{[v(0)]^2} = z(0) = \frac{kT}{m} \qquad (2.10)$$

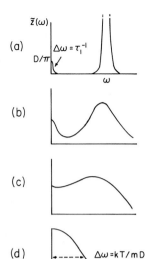

Fig. 2.4a–d. Sketch of the spectral density $\tilde{z}(\omega)$ from (a) an idealized solid with jump diffusion through (b) and (c) liquids to (d) Brownian motion ("free" diffusion) in a dense gas [2.12]. $\Delta\omega = \tau^{-1} = kT/mD$

where m is the mass of the ion. To the extent that one can neglect correlation effects in an ionic conductor, $\tilde{z}(\omega)$ is proportional to the spectral density of the current correlation function and hence to the frequency dependent ionic conductivity

$$\sigma(\omega) = \frac{\pi n(Ze)^2}{kT}\, \tilde{z}(\omega). \tag{2.11}$$

In Fig. 2.4, qualitative responses of $\sigma(\omega)$ or $\tilde{z}(\omega)$ for an idealized solid and for liquids are shown for reference [2.2].

It is instructive for the above reason to derive $\tilde{z}(\omega)$ deductively with a certain model and compare the results with measured frequency response of the conductivity $\sigma(\omega)$ of different superionic conductors based on (2.11) [2.12, 49–51]. Among these, *Zeller* et al. [2.50–52] discussed the problem based on the memory function formalism. First, let us look at the highly nonlinear Langevin equation which describes the motion of particles in the periodic potential:

$$m\ddot{x} + m\Gamma\dot{x} + f(x) = K \tag{2.12}$$

where m is the mass of the particle, Γ the damping, $f(x)$ the restoring force and K the stochastic force. It is easily visualized for physical reasons that this equation asymptotically describes a damped harmonic oscillator at high frequencies and a diffusion at $\omega \to 0$. $f(x)$ in (2.12) may be replaced by a memory function $M(t)$ which provides the correct asymptotic properties

$$m\ddot{x} + m\Gamma\dot{x} + m\omega^2 \int_0^t M(t-t')\dot{x}(t')dt' = K. \tag{2.13}$$

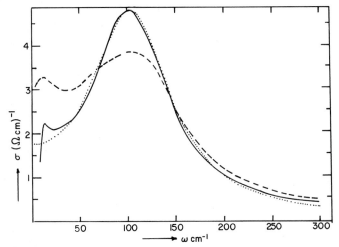

Fig. 2.5. Frequency dependent conductivity of α-AgI ($T=453$ K, solid line), liquid AgI ($T=853$ K, dashed line) and fit of (2.15) to α-AgI (dotted line). The fit parameters are: $\omega_0 = 105$ cm^{-1}, $\Gamma = 45$ cm^{-1}, $\gamma = 53$ cm^{-1} [2.52]

The simplest $M(t)$ which leads to diffusion as $\omega \to 0$ and to oscillation at high ω is

$$M(t) = e^{-\gamma t}. \tag{2.14}$$

Equation (2.13) then directly leads, to a good approximation [2.50], to

$$\sigma(\omega) = \frac{n(Ze)^2}{m} \cdot \frac{1}{-i\omega + \Gamma + \dfrac{\omega_0^2}{-i\omega + \gamma}} \tag{2.15}$$

where ω_0 is the resonance frequency. Thus substances are characterized by three parameters ω_0, Γ, and γ. A fit to $\sigma(\omega)$ of α-AgI is shown in Fig. 2.5. A further improvement of the theory can be made, for example, by taking into account in the model the effect of the lattice polarizability and correlation effects [2.52].

Under an identical structural condition, the location of the conductivity maximum ω_0 should then scale with the inverse of the square root of ionic carrier mass [2.52]. In β-alumina, ionic carriers can be readily exchanged from Na to numerous ions from molten salts [2.53–55]. *Allen* and *Remeika* [2.56] carried out experiments with ion exchanged β-alumina and located the maximum corresponding to ω_0 in the infrared region. The result indicates that ω_0 does not necessarily scale with the inverse of the square root of ionic mass. For example, K − β-alumina was found to have a higher frequency than Na − β-alumina. An explanation for this result was attempted by *Wang* et al. [2.57] (see Sect. 2.3).

If an ion takes a long time to move from one equilibrium site to another, it would be worthwhile to investigate the probable path which ions should follow between the two sites in crystallographic channels. *Flygare* and *Huggins* [2.58, 59] calculated such a possible path under a static condition for α-AgI. The energy E of an arbitrarily taken mobile ion (i-th ion) in a crystallographic channel is given by

$$E = e^2 \sum_j \frac{q_i q_j}{r_{ij}} - \frac{e^2}{2} \sum_j \frac{\alpha_j q_i}{r_{ij}^4} + \sum_j \beta_{ij} e^{(r_i + r_j - r_{ij})/\varrho}. \tag{2.16}$$

The sum over j is over all lattice ions, q_i and q_j are the charge of the mobile ion and that of the j-th fixed lattice ion, respectively, α_j is the dipolar polarizability of the j-th fixed lattice ion (I^-) and r_{ij} is the distance from the mobile ion to the j-th lattice ion. The second term in (2.16) is the polarization self-energy of a nonpolarizable (or assumed to be negligibly small) mobile cation in the lattice of fixed polarizable anions (I^-). Here the orientational dipole-dipole terms are neglected. The third term is the overlap repulsion between closed shell ions with parameters β_{ij} and ϱ to be determined depending on the nature of the interacting ions [2.60, 61]. r_i and r_j are the ionic radii of ions i and j. Equation (2.16) is valid only for $r_i + r_j < r_{ij}$, but because the crystal is loose, (2.16) is sufficient to find the minimum of the energy under this condition avoiding the difficulty of divergence as $r_{ij} \rightarrow 0$. In addition, it is necessary to account for the presence of the average positive charge distribution caused by the assembly of mobile Ag^+ ion distribution. Along with the first term in (2.16), this energy term represents the Madelung energy. But it is found that terms sensitive to the location of the mobile ion in the channel are the second and the third term while the Madelung term is practically insensitive to the location of the mobile ion. Therefore, the minimum energy path is determined essentially by the balance of the second and the third term. The energy is thus calculated as a function of cation position (and size) along a tunnel parallel to the cube edge in the x direction passing through the point $x = 0$, $y = 0$, and $z = a/2$ allowing the motion of ions in both y and z directions.

From the functional form of the second and the third term, it is obvious that the path and the energy depend sensitively on cation size. In the following, qualitative conclusions are summarized.

1) In all cases, the minimum energy path deviates considerably from the centerline of the tunnel.

2) A cation moving down the tunnel passes between pairs of anions with center lines alternately rotated 90°. In the case of small anions ($r < 0.8$ Å) the polarization term in (2.16) dominates the potential energy causing the cation minimum energy path to split into two equivalent paths which deviate toward the alternating anion pairs. On the other hand, the repulsive terms are more important for larger cations ($r > 0.8$ Å) causing path deviations in a plane normal to the anion-anion pair axes. These effects are shown in Fig. 2.6.

3) By finding the minimum energy path in the x-direction, the variation of potential energy during translation through the tunnel can be obtained. The difference between the maxima and the minima in energy along x corresponds to

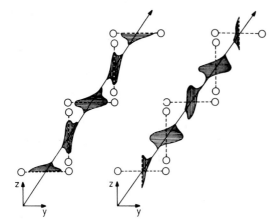

Fig. 2.6. Schematic drawing of loci of minimum energy path for cations of different sizes; on the left, small ions are attracted toward pairs of I⁻ ions (indicated by circles), on the right, paths of larger ions are orthogonal to the anion-anion axis [2.58]

the activation energy of motion of cations. A minimum in the activation energy is found for intermediate size ions.

Although the theoretical developments are as yet crude, the results obtained seem to explain many experimental observations. In the first place, the minimum energy path calculated is consistent with the results of neutron diffraction concerning the density of distribution of Ag ions in α-AgI [2.40–42]. Second, the result shows that the conductivity strongly depends on cation size and, at a favorable size, the activation energy of motion can be extremely low. In ion-exchanged β-alumina, *Yao* and *Kummer* [2.53] actually found that the activation energy of diffusion for Li was higher than Na$^+$ or Ag$^+$ irrespective of its small size, corroborating the calculated results.

The above calculation indicates the importance of the lattice polarizability for fast ionic conduction although the effect is indirect. Here, the calculation is made in a static fashion. However, the importance of the lattice polarizability may be connected to lattice dynamics for moving ions. An ion will either attract or repel its neighbors from the equilibrium positions, thus creating a local polarization cloud about the central ion. When the ion moves, it must carry this deformation cloud with it. Therefore, the ion motion and the lattice motion are connected. The theory of this sort was first developed for electron motion, and is called "small polaron theory" [2.62]. *Flynn* and *Stoneham* [2.3, 63] suggested that the same theory could apply to a diffusing ion, and this may be called the "ionic polaron". Assuming that the local polarizations are linearly coupled to the ion, *Mahan* and *Pardee* [2.64] have shown that the conductivity has an activation energy of

$$u_p = \frac{Z^2 e^2}{\pi d}\left(\frac{1}{\varepsilon_\infty} - \frac{1}{\varepsilon_0}\right) \tag{2.17}$$

where d is the jump distance, Z is the valence, and ε_0 and ε_∞ are the static and dynamic dielectric constants. Thus a large amount of local polarization about

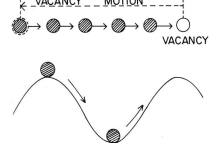

VACANCY _ _ MOTION _ _ _ _ _

VACANCY

Fig. 2.7. Caterpillar mechanism of ionic motion

Fig. 2.8. Cooperative motion of two ions with zero activation energy

the diffusing ion will impede its motion. It seems, however, that the polaron effect is, except possibly for hydrogen diffusion, small compared to the static activation energy for classical systems and should be only partly responsible for the activation energy of ionic motion.

2.3 Cooperative Motions of Ions

The common description of ionic diffusion or conduction is based on one ion hopping in the lattice in terms of the random walk theory. It is, however, very difficult to give reasonable explanations of extremely fast conduction with extremely low activation energies within the framework of the random walk theory. To reconcile these difficulties, some models of cooperative motion of ions have been proposed.

One of the most commonly known cooperative motions of ions adopted in the theory of diffusion is the interstitialcy motion. Here, two ions, the one on an interstitial site and the other on an adjacent regular site, cooperate in such a way that the interstitial ion replaces the ion on the regular site while forcing the ion on the regular site out into an adjacent interstitial site [2.2]. A similar model based on a vacancy mechanism which involves a larger number of ions has also been proposed in connection with high ionic conductors. In this model, a row of ions moves cooperatively by one atomic distance each through a vacancy so that, as a result, the vacancy moves a long distance nd, where n is the number of ions involved in the row and d is the atomic distance (Fig. 2.7). It is apparent that such a cooperative motion, called the caterpillar mechanism [2.65] has a distinct advantage of increasing the jump distance if it occurs.

Cooperative motions of ions like those discussed above are found especially effective in disordered lattices in reducing the activation energy of motion of the group of ions. In disordered lattices, due to the availability of extra interstitial sites, the locations of all the conducting ions are not necessarily in the regular periodic potential wells of the lattice. Therefore, in a cooperative motion of ions with a fixed interionic distance, all the ions do not follow the path in phase with the period of the potential. In Fig. 2.8, the relation is shown schematically for two

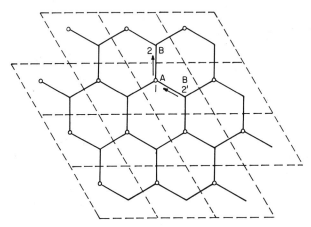

Fig. 2.9. The two dimensional honeycomb lattice indicating the Beevers-Ross sites (A) and the anti Beevers-Ross sites (B) of β-alumina [2.57]

ions where one ion sits in a potential well while the neighboring ion sits on top of the potential barrier. If these two ions move keeping the fixed interionic distance, a situation arises in which one ion is moving up the potential barrier while the other ion is moving down. As a result, the height of the potential barrier for the cooperative motion is cancelled out.

A theory utilizing a mechanism similar to the above has been applied to β-alumina by *Wang* et al. [2.57]. Essential features of the theory can be explained with the aid of the illustration of the conducting plane of β-alumina in Fig. 2.9. The dotted lines show the oxygen lattice whereas the two dimensional honeycomb lattice indicates the sites for Na^+ ions (or for other monovalent ions M^+ in ion-exchanged β-alumina) according to the Beevers-Ross structure. The two dimensional honeycomb lattice can be divided into two sublattices A and B. A is called the Beevers-Ross (B–R) site whereas B is called the anti B–R site. In an idealized stoichiometric case $(Na_2O \cdot 11Al_2O_3)$, A sites are the sites for Na^+ (or M^+) ions. The midpoints between neighboring A and B sites are called the mid-oxygen sites. For calculation, the "stoichiometric composition" is first assumed and all the conducting M^+ ions are on the B–R sites. In this structure, the variation of the total energy is then calculated as a function of the location of one M^+ ion (on Site 1) as it is moved along the straight line toward the nearest anti B–R site (Site 2) while all other ions are fixed. Such calculations show that the total potential energy increases by approximately 2 eV for any M^+ ion. In other words, this proves that the anti B–R sites are not the stable site for M^+ ions. Next, one M^+ ion is added to an anti B–R site (Site 2′). As an M^+ ion at one of the B–R sites nearest to the extra M^+ ion is again moved along the straight line toward a nearest vacant anti B–R site (Site 2), the variation of the total crystal potential energy is calculated, while a number of other M^+ ions are allowed to adjust their positions such that the total potential energy is minimized. The calculation shows that the extra M^+ ion which resides at the anti B–R site is

Table 2.2. Calculated vibrational frequency along the trajectory (*Wang* et al. [2.57]

Ion	v_{pair} [cm^{-1}] (v_0)	v_{single} [cm^{-1}]
Na$^+$	39.0	89.7
Ag$^+$	16.1	41.9
K$^+$	52.0	97.7
Rb$^+$	50.4	78.2

found to be unstable and the minimum potential energy corresponds to the paired motion of the extra M$^+$ ion with another M$^+$ ion being moved as shown in Fig. 2.9. A minimum in the potential energy is found when both the ions are located at the mid-oxygen sites. The potential barriers for the in-phase motion of the pair are found to be close to the activation energy of motion obtained experimentally (0.165 eV for Na$^+$), although the value depends on the number of other M$^+$ ions whose positions are adjusted. The situation is similar to that indicated in Fig. 2.8 and the calculation shows fully the advantage of the cooperative motion of ions in a disordered lattice. The results thus strongly suggest an interstitialcy mechanism of diffusion in β-alumina as long as the number of extra M$^+$ ions is small compared to the number of the anti B–R sites (or of the B–R sites) as the nature of the calculation indicates.

If the number of extra M$^+$ ions is small enough and if the interaction between individual pairs is neglected, the ionic conductivity due to the in-phase motion of the pairs or to the interstitialcy mechanism can be calculated utilizing the random walk theory [2.1–3]. If the concentration of the excess M$^+$ ions is n/N (n is the number of excess M$^+$ ions and N is the number of the B–R sites), the jump frequency of the ionic pair is given by

$$v = 2v_0(1 - n/N)\exp(-u_a/kT) \tag{2.18}$$

where v_0 and u_a are the oscillation frequency and the potential energy barrier, respectively, of the in-phase motion of the two particles. Then the conductivity is given by

$$\sigma = \sigma_0 \exp(u_a/kT) \tag{2.19}$$

where

$$\sigma_0 = (e^2 d^2/kT)v_0 n(1 - n/N). \tag{2.20}$$

The value v_0 can be evaluated from the calculated potential energy curves for the in-phase motion of two particles and is given in Table 2.2.

Equation (2.19) and (2.20) with 16% excess M$^+$ ions are found to give a reasonably good agreement with experimental results for Na$^+$ and Ag$^+$. The calculated values of v_0 for different M^+ ions are also found to agree reasonably well with experiments.

The above calculation was meant to indicate that only the interstitialcy motion of ions is plausible in β-alumina in view of its small activation energy of diffusion. In this respect, a comparison of β- and β″-alumina may be interesting for the following reason.

Compared to β-alumina, β″-alumina has a somewhat better ionic conductivity but with a similar activation energy. Here, the B–R sites and the anti B–R sites are equivalent and the overall concentration of M^+ ions with respect to the B–R sites and the anti B–R sites combined is approximately the same as β-alumina. In β″-alumina, therefore, it is apparent that the concept of the in-phase motion of pairs to lower the activation energy is not applicable because all the available sites are equivalent. If so, the conduction mechanism in β″-alumina must be that of one ion hopping. A calculation would thus be of interest to see whether the activation energy of motion of one ion hopping in β″-alumina can be as small as that of in-phase motion of pairs in β-alumina. In fact, in a disordered system in which a large number of available sites exist, no clear distinction between the vacancy mechanism, interstitial mechanism or interstitialcy mechanism can be made. (See Sect. 2.4.)

Another type of cooperative motion of ions often discussed in fast ionic conductors so far is a domain wall motion [2.66, 67]. The possibility of the existence of domains in β-alumina type compounds was first pointed out by *Roth* [2.68] in connection with the nonstoichiometry of this type of compounds. In the zeroth approximation, the energy of a domain wall does not depend on the location and a high mobility of domain walls is connected to this fact. Domain wall motion is especially well known in ferromagnetic substances [2.69]. Domains exist here to reduce the overall magnetostatic energy, and, by the application of a weak external magnetic field, domain walls move by a substantial distance to achieve the magnetization process. The analysis of such domain wall motions in a variety of materials reveals that, in order for characteristics of materials to be largely due to domain wall motions, certain general conditions must be satisfied.

1) Each set of domains must be physically equivalent in the absence of the external force or field.

2) One set of domains should be definitely favored over the others upon the application of external force or field so that one set of domains grows at the expense of the others.

3) Domains should exist under a (quasi) equilibrium condition so that domain structures should be maintained or reestablished after the removal of the external field.

The concept of domains in fast ionic conductors will be examined in the light of the above statements.

Domains discussed so far in fast ionic conductors are essentially those of antiphase domains [2.66–68]. As an idealized case, let us examine a β″-alumina type material where the B–R sites and the anti B–R sites are equivalent (Fig. 2.2). Then these two kinds of sites form two equivalent sublattices of the two dimensional honeycomb lattice A and B, respectively. If there exist repulsive interactions among conducting ions and the concentration of ions is around 0.5 with respect to the total number of sites, these ions tend to form an ordered structure consisting of ions and vacancies [2.21, 70]. The ordering thus creates preferential sites for ions among otherwise equivalent sites and the two sublattices are now distinguishable. The ordered structure can, however, nucleate in either of the two sublattices so that two sets of antiphase domains can be established.

There are, however, unfavorable conditions to expect a large contribution from domain wall motion in such a material. In the first place, the domain walls thus formed have charges and hence the domain wall energy is high [2.71]. This makes the formation of domain walls unfavorable. Unlike ferroelectrics where domains exist in order to reduce the overall electrostatic energy, in fast ionic conductors, there is no need energetically to establish domain structures, and domains can exist only as a metastable state. In addition, it is known that such domains can exist as a metastable state only if at least four sets of domains meet at a corner [2.23]. Two sets of antiphase domains which are likely to occur in β''-alumina-type structures are known to be unstable against the disturbance, and the domains tend to vanish by heat treatment or under the influence of an external field [2.23]. Even if domains exist, neither set of domains is preferred under an external electric field so that one set of domains does not grow at the expense of the other.

Domain structures discussed so far for fast ionic conductors, however, are not for β''-alumina but for β-alumina (like Ag–β-alumina where some Ag^+ ions exist on the anti B–R sites [2.66–68]). Because the B–R sites and the anti B–R sites are not equivalent in β-alumina, the two antiphase domains discussed are not even equivalent and there is no reason to expect antiphase domain structures of the sort discussed above in the case of β-alumina. In short, it is unlikely to observe antiphase domains and to expect a large contribution to the ionic conductivity from domain wall motion in β-alumina-type compounds. Contribution of domain wall motion to fast ionic conduction is also suggested in other fast ionic conductors [2.72]. There is, however, no experimental evidence at the moment to confirm the contribution of domain walls to the conductivity.

It might be necessary here to cite an interesting model for fast ionic conductors proposed by *Rice* and *Roth* [2.73] which resembles closely that of semiconductors. The model is based on the hypothesis that there exists in the ionic conductor an energy gap ε_0 above which ions of mass M, belonging to the conducting species, can be thermally excited from localized ionic states to free-ion-like states in which an ion propagates throughout the solid with a velocity v_m and energy $\varepsilon_m = (1/2) M v_m^2$. To account for the interaction with the rest of the solid, a finite lifetime τ_m for such an excited free-ion-like state is further introduced. Then, based on the Boltzmann transport equation, simple expressions can be derived for the ionic conductivity σ, ionic thermal conductivity

K_1, thermoelectric power Q_1, etc., as well as the frequency dependence of the conductivity $\sigma(\omega)$. For example, in the low temperature limit in which $\varepsilon_0/kT \gg 1$, the result for the ionic conductivity becomes

$$\sigma = (1/3)(Ze)^2/(kT)nv_0l_0 \exp(-\varepsilon_0/kT) \tag{2.21}$$

where n is the density of potentially mobile ions and $\varepsilon_0, v_0, l_0 = v_0\tau_0$ characterize the free-ion-like state at the energy gap ε_0. This expression is to be compared with that of the usual hopping model

$$\sigma = (1/3)(Ze)^2/(kT)nd^2v_0 \exp(-u/kT). \tag{2.22}$$

Both equations become identical if ε_0 is identified with u, the mean free path l_0 with d and the inverse lifetime $1/\tau_0$ with v_0. Because of the relation

$$v_0 = \sqrt{2\varepsilon_0/M} \tag{2.23}$$

σ in (2.21) is determined by two adjustable parameters ε_0 and τ_0 which can be determined empirically with the knowledge of Z, n, and M. It is claimed that l_0 and v_0 for this approach have clearer physical meaning than obscure quantities like d and v_0 in terms of the hopping model in the case of liquid-like motion in fast ionic conductors.

A characteristic of the Rice-Roth model is the assumption of the free-ion-like relation

$$\varepsilon_m = (1/2)Mv_m^2. \tag{2.24}$$

In view of the identification

$$v_0 = 1/\tau_0 \tag{2.25}$$

the relation

$$v_0 = (1/l_0)\sqrt{2\varepsilon_0/M} \tag{2.26}$$

is derived. In the case of atomic self-diffusion where one can identify

$$l_0 = d \tag{2.27a}$$

and

$$\varepsilon_0 = u \tag{2.27b}$$

(2.26) leads to the value of v_0 which coincides with the observed Debye frequencies. This they consider to constitute fairly satisfactory evidence for the basic validity of the free ion-like relation between ε_m and v_m.

Empirical analysis of the conductivity of fast ionic conductors in terms of (2.21) requires unusually large values of l_0 of the order of 1000 Å for certain kind of substances [2.73]. Although this feature is very attractive empirically to obtain large enough prefactors, it is, at the same time, very puzzling physically. Different from the case of electrons, the space where ions can move is quite limited and the correlation among conducting ions is also quite strong. Unless most of the lattice sites available for conducting ions are vacant, it is difficult to accept the value of l_0 of the order of 1000 Å literally. *Haas* [2.74] has pointed out that the characteristic relation of the model, (2.24), can be derived for a simple harmonic oscillator hopping-type model and questioned the necessity for involving a free-ion-like model. It seems necessary yet to examine the physical basis of the model in more detail in order to rationalize the use of the free-ion-like model.

2.4 Ionic Diffusion and Conduction in Disordered Systems

In the previous sections, features of crystallographic disorders related to fast ionic conductors and ionic motions involved in such disordered systems are explained. To understand features of ionic transport in such systems fully, however, a quantitative description of ionic conductivity equivalent to the random walk theory of ionic conduction and diffusion in nearly perfect solids is desirable. A major feature of the disorder which affects the ionic conduction in fast ionic conductors is that the number of available sites is greater than that of conduction ions. In such systems, mutual interactions among ions strongly affect the diffusion. The problem is then a many body problem very similar to that of electron conduction in metals. The space available for ion transport is, however, much more limited than in the case of electrons and the correlations among ions are much stronger. Therefore, the techniques developed in dealing with electronic conduction are not readily applicable. *Sato* and *Kikuchi* [2.75–77] applied a technique utilized in dealing with the Ising problem to take this situation into account and we describe this treatment next in some detail.

To deal with these essential features of the ionic transport in disordered systems, the following simplifying assumptions are introduced [2.21]. We assume that the species of ions which does not contribute to ionic conduction forms a rigid lattice. Mutual interactions are limited to conducting ions and these are distributed among potential wells whose number is greater than that of the conducting ions. These conducting ions are assumed either to stay in such potential wells or to be allowed to move to neighboring sites if they are vacant. In addition, a Markovian process is assumed for the motion of ions such that a jump of an ion from one site to another is independent of its previous history. The latter assumption is well represented in the case of hopping diffusion. However, in dealing with the case of low frequency limit of conductivity ($\omega \rightarrow 0$), the assumption seems to hold approximately even in fast ionic conductors where the dwell time becomes of the same order as the hopping time, although a description of the dynamical response of the system or a calculation of $\sigma(\omega)$ has to be sacrificed.

The calculations of diffusion and ionic conduction are carried out in the following fashion. Isotope diffusion or ionic conduction in the crystal is essentially a time dependent Ising problem (or the problem of the order-disorder transformation in alloys). In the stationary state where the rate of flow of ions does not change with time, the distribution of ions (including isotope ions) and vacancies on the lattice sites also does not change. Because of the presence of the chemical potential gradient, however, the distribution of ions deviates from that of equilibrium. This deviation creates a flow of ions along the chemical potential gradient but this occurs in the stationary state in a way not to disturb the overall distribution of ions. By calculating the flow rate under this condition, one can obtain the ionic conductivity or the diffusion coefficient of both ordinary ions and isotope ions and, hence, the correlation factor. The equilibrium distribution of ions is calculated by the cluster variation method [2.22, 23, 78], a closed formula approximation of the Ising problem. The deviation from the equilibrium value and the change of state or the rate of flow of ions are determined by the path probability method [2.79] which is an extension of the cluster variation method to the nonequilibrium case.

The theory basically deals with the vacancy mechanism of diffusion. However, in disordered systems, no clear distinction can be made between the vacancy mechanism and the interstitial or interstitialcy mechanism. The results also include features of cooperative motion of diffusing ions. For example the existence of available sites more numerous than the number of conducting ions will lead to a concept of an effective number of contributing ions which is different from either the number of ions or that of vacancies. Also effects of preferential sites among available sites on diffusion will be clarified. Further, repulsive interactions among ions will lead to a reduction in the activation energy of motion and in the preexponential factor and to a possible ordered distribution of ions among available sites. The concepts of the correlation factor of isotope diffusion [2.80] and of the Nernst-Einstein relation which are basically developed based on the random walk theory of diffusion will also be reexamined.

In the following, a brief description of the theory based on two idealized lattice models and a qualitative explanation of the results will be given. The details of the derivation are found in the original papers [2.21, 74].

The calculation will be applied to the two dimensional honeycomb lattice shown in Fig. 2.2. As has been discussed several times in previous sections, the two dimensional honeycomb lattice represents the conducting layer of both β- and β″-alumina with Na⁺ the conducting cations. In β″-alumina, both the A and B sites are equivalent while in β-alumina, the two sublattices are distinguishable. The mid-oxygen sites defined in Sect. 2.3 are totally disregarded in the present calculation. Nevertheless the two lattice models defined here are simply referred to as β″-alumina and β-alumina hereafter, because the models still closely resemble these two materials. (See also Chap. 5.)

In the calculation, conducting ions (referred to as Na^+ ions hereafter) are assumed to occupy the lattice sites partially, and the concentration of ions with respect to the total lattice sites is given by ϱ. ϱ is assumed to take any value between 0 and 1 in the calculation so that the effect of the density of conducting ions can be investigated. An interaction potential ε is assumed for a pair of nearest neighboring conducting ions. The sign of ε is chosen so that $\varepsilon < 0$ corresponds to repulsion and $\varepsilon > 0$ attraction between ions. The formulation allows both positive and negative values of ε. ε is supposed to represent all possible interactions and neither the origin of ε nor the evaluation of it in real materials is sought here. The difference in the occupancy of ions between the B–R sites (A) and the anti B–R sites (B) is represented by the activation energy w in β-alumina so that the effect of the repopulation of ions among various sites along with the effect of vacant sites can be investigated here. Because the case of β″-alumina is simpler, the mathematical steps will be explained only for β″-alumina.

2.4.1 The Path Probability Method

The State Variables

To specify the state of the system, a certain number of variables called the state variables are defined first according to the scheme of the cluster variation method (see Table 2.3).

At a fixed value of ϱ, a certain lattice site either is occupied by an ion or is vacant. The fraction of lattice sites which are occupied by Na^+ ions is written as p_1 and the fraction of vacant lattice sites is written as p_3. Here 1 and 3 stand for Na^+ and vacancy, respectively. p_1 and p_3 are the state variables related to a lattice site. For a pair of lattice sites, the possible configurations are $Na^+ - Na^+$, $Na^+ - v$, and $v - v$ where Na^+ represents an occupied site and v represents a vacant site; the fractions of these configurations are written as y_{11}, y_{13}, and y_{33}, respectively. These y's are also the state variables related to a pair of sites. The procedure can be continued for larger clusters of lattice points and we can conceive a hierarchy of a large number of state variables. These variables are used as variation variables to evaluate the total free energy of the system in the cluster variation method. In actual formulations, however, we use a limited number of them to make mathematical analysis tractable, although the derived results are approximate. The formulation in which state variables for a pair and a point (site) only are used is called the pair approximation. Likewise, the point approximation can be defined. The point approximation corresponds to the Bragg-Williams approximation. We limit ourselves to the pair approximation in the present formulation.

The significance of the state variables in statistical mechanics is understood as follows. The set of state variables which may be written as $\{y\}$ for short specifies the state of the system. When the set $\{y\}$ is defined, we can formulate the probability Prob$\{y\}$ that the system takes a state $\{y\}$ when the system is in contact with a heat bath, or when the system is kept isolated. The equilibrium

Table 2.3. State variables of β″-alumina

(a) The letter t indicates that $p_i^{(v)}$ depends on time

Configuration of a v-th lattice point	State variable
Na$^+$	$p_1^{(v)}(t)$
Na*	$p_2^{(v)}(t)$
Vacant	$p_3^{(v)}(t)$

(b) $i = 1, 2,$ and 3 indicate Na$^+$, Na*, and v, respectively

Configuration	State variable
on a v-th bond	
i———j	$x_{ij}^{(v)}(t)$
v-th $(v+1)$-th	
i———j	$y_{ij}^{(n)}(t)$

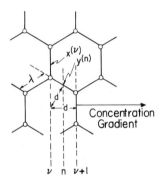

Fig. 2.10. Two dimensional honeycomb lattice representing the conducting layer of β″-alumina. v and $v+1$ indicate the position of lattice points, while n indicates a bond connecting v-th and $(v+1)$-th points [2.21]. λ indicates the nearest neighbor distance

state of the system is derived as the state $\{y_e\}$ which maximizes Prob$\{y\}$. This is the procedure of the cluster-variation method in equilibrium statistical mechanics, and the procedures here are equivalent to those utilized in treating the problem of the order-disorder transformations in alloys.

When a concentration gradient is imposed on the system, as shown in Fig. 2.10, the state variables p's and y's become dependent on how far down the lattice site is located along the concentration gradient, and are defined with a suffix v as in Table 2.3. It may be noted in Fig. 2.10 and Table 2.3(b) that two kinds of bonds specified by x_{ij} and y_{ij} are distinguished with respect to the direction of the concentration gradient. In Table 2.3, we also include Na* which indicates an isotope of Na$^+$ to be used in formulating the isotope diffusion coefficient.

In the pair approximation, those listed in Table 2.3 are all the state variables that are needed to specify the state of the system at time t. When the state of the system changes in time, values of these variables change. These changes are described by the path probability method explained in subsequent sections.

The Path Variables

Generalizing the state variables, we introduce the concept of the path variables which describe how the state of the system changes. It is of the form of $P_{i,j}^{(v)}$ $(t, t + \Delta t)$ in which i is the configuration of a v-th lattice site at time t and j is the configuration of the same lattice site at $t + \Delta t$. $P_{i,j}$ represents the probability that such a change occurs at the lattice site, i and j represent 1, 2, or 3 and stand for species Na^+, Na^*, or v, respectively.

In the same sense, we can define $Y_{23,32}^{(n)}$ $(t, t + \Delta t)$. This is the probability of finding such a lattice site at the n-th location that the configuration is $Na^* - v$ at t and $v - Na^*$ at $t + \Delta t$. Note that 2 stands for Na^* just as 1 and 3 stand for Na^+ and v, respectively. Because $Y_{23,32}^{(n)}$ represents migration of a Na^* towards the plus direction (direction of the concentration gradient) through a bond, we may write it as $Y_{+2}^{(n)}$ for short. Similarly, we can define $Y_{-2}^{(n)}$. The difference between the two is the net probability that a Na^* ion moves towards the plus direction in Δt:

$$\Phi_2^{(n)} \equiv Y_{+2}^{(n)} - Y_{-2}^{(n)}. \tag{2.28}$$

When the systems is in equilibrium, Φ_2 vanishes. When a small concentration gradient $dp_2^{(v)}/dv$ of Na^* ions is imposed on the system, Φ_2 does not vanish but is proportional to the concentration gradient. The diffusion coefficient of Na^*, D_2, is defined as the ratio of Φ_2 to the gradient, except for the normalization constant c:

$$c\Phi_2^{(n)} = -D_2 \frac{dp_2^{(v)}}{dv}. \tag{2.29}$$

The theory of the diffusion coefficient is to find the relation between $\Phi_2^{(n)}$ and $dp_2^{(v)}/dv$. To accomplish this there are two tasks to be done. They are described in the subsequent two sections.

The Path Probability

Suppose the system is in a state specified by a set of the state variables defined in Table 2.3. Let us write them schematically as $\{y(t)\}$. In general, this state is neither the state in equilibrium nor the state in the steady state. It is necessary to know how the system changes starting from the given state $\{y(t)\}$, particularly how $Y_{\pm 2}$ of the previous section should depend on $\{y(t)\}$.

This problem cannot be answered within the framework of equilibrium statistical mechanics. However, it can be answered by using probabilistic reasoning when the unit kinetic process is assumed given. This formalism is called the "path probability method".

When the state of the system $\{y(t)\}$ is given at t, it is *kinetically* possible that the system changes into any one of many different states $\{y'\}$ at $t + \Delta t$. Suppose we can formulate the probability of changing into the state $\{y(t + \Delta t)\}$; we write it as $\mathscr{P}[\{y(t)\}, \{y(t+\varDelta t)\}]$. (The function $\mathscr{P}$ thus has meaning similar to the partition function in equilibrium statistical mechanics.) Then the state into which the system most probably changes is $\{\hat{y}(t+\varDelta t)\}$, which makes $\mathscr{P}$ a maximum for the fixed $\{y(t)\}$.

The problem is then how to write the path probability function $\mathscr{P}[\{y(t)\}, \{y(t+\varDelta t)\}]$. It can be done by extending the conventional statistical mechanics into the time dependent regime by treating the time axis as though it were the fourth space axis. For our present purposes it is sufficient to say that the most natural way of writing the path probability function $\mathscr{P}$ is in terms of the path variables $\{Y(t, t+\varDelta t)\}$ introduced in the previous section, rather than in terms of the state variables $\{y(t)\}$ and $\{y(t+\varDelta t)\}$. The writing of $\mathscr{P}$ in terms of path variables can be made in analogy to the construction of the partition function by the cluster variation method in terms of state variables.

After $\mathscr{P}$ is written in terms of $\{Y(t, t+\varDelta t)\}$, we derive the natural change of the system by maximizing $\mathscr{P}$ with respect to $\{Y(t, t+\varDelta t)\}$, keeping $\{y(t)\}$ fixed. The maximization in the pair approximation leads to the expression of $\{Y(t, t+\varDelta t)\}$ written in terms of $\{y(t)\}$. The result of our concern is

$$Y_{+2}^{(n)}(t, t+\varDelta t)=(\varDelta t)\theta e^{-u/kT}W_{+2}^{(n)}(t)y_{23}^{(n)}(t)$$
$$Y_{-2}^{(n)}(t, t+\varDelta t)=(\varDelta t)\theta e^{-u/kT}W_{-2}^{(n)}(t)y_{32}^{(n)}(t). \tag{2.30}$$

The factor $\theta \exp(-u/kT)$ is the probability of a unit jump, θ being the vibrational contribution to the jump frequency, or the attempt frequency which is specified as w_0 earlier, and u is the activation energy for a jump of a Na* ion into a vacant site. $W_{\pm 2}^{(n)}$ represents the effects of the bond-breaking of the jumping Na* ion and is written in terms of $y^{(n-1)}$, $y^{(n)}$, and $y^{(n+1)}$. It is important to note here that, in this approximation, the most probable path is given by a "so-called" superposition principle in a way that W is given by the state immediately before the ion jumps.

Substitution of (2.30) in (2.28) allows us to write the flux $\varPhi_2^{(n)}$ in terms of $y^{(n-1)}$, $y^{(n)}$, and $y^{(n+1)}$ at time t. This is the first task of the two.

Stationary State Condition

The expression (2.30) holds for any state given at time t. On the other hand, the diffusion coefficient is defined for a stationary state. Therefore, the required second task is to find the state variables which represent the most probable state in the stationary state under the imposed concentration gradient.

Although the stationary state can be represented in many different ways, one of them is

$$y_{23}^{(n)}(t+\varDelta t)-y_{23}^{(n)}(t)=0 \tag{2.31}$$

which simply indicates that the nearest neighbor relation between Na* and v does not change with time. Both $y_{23}^{(n)}(t+\varDelta t)$ and $y_{23}^{(n)}(t)$ can be written in terms of the path variables $\{Y(t, t + \Delta t)\}$, which in turn are expressed in terms of the state variables $\{y(t)\}$ at time t. Thus, (2.31) is an equation to determine the set of the state variables in the stationary state.

2.4.2 β″-Alumina

The Tracer Diffusion Coefficient

When the concentration gradient $dp_2^{(v)}/dv$ is not large, we can expand $\Phi_2^{(n)}$ in (2.28) and the left-hand side of (2.31) in powers of $dp_2^{(v)}/dv$ and keep only the linear terms. This procedure is consistent with the definition of the diffusion coefficient in (2.29).

The final expression for the diffusion coefficient of Na* in β″-alumina thus obtained is written as follows:

$$D_2 = d^2 \theta e^{-u/kT} V W f \tag{2.32}$$

where d is the jump distance along the concentration gradient and is defined in Fig. 2.10. Note here that (2.32) is identical with the random walk approach except for the factors V, W, and f. These factors are necessary to take into account the many body effect arising from the large number of vacancies and from the mutual interaction among ions. θ is the attempt frequency of an ion if there were no effect from the change of the surroundings.

The factor V is called the vacancy availability factor and is the probability of finding a vacancy next to a Na* ion. Hence, this eventually gives the *effective* number of vacancies. It is written as

$$V = y_{23e}/p_{2e} \tag{2.33}$$

where the subscript e indicates the equilibrium value. W indicates the effect of the surroundings on the jump frequency and is the equilibrium value of $W_{\pm 2}^{(n)}$ in (2.30). It is written as

$$W = [(y_{12e} + y_{22e}) e^{-\varepsilon/kT} + y_{32e}]^2 p_{2e}^{-2} \tag{2.34}$$

where ε is the pairwise interaction energy between two Na$^+$ ions on adjacent lattice sites. The Boltzmann factor $\exp(-\varepsilon/kT)$ represents the extra activation energy needed to break a Na$^+$ − Na$^+$ bond. f in (2.32) represents the efficiency of a jump in contributing to the diffusion coefficient and corresponds to the correlation factor in the isotope diffusion [2.4, 21].

The three factors V, W, and f in (2.32) depend strongly on the sign of ε. When $\varepsilon = 0$, i.e., when there is no interaction among Na$^+$ ions,

$$V = 1 - \varrho. \tag{2.35}$$

Equation (2.35) express that Na$^+$ ions exist randomly within the layer. When $\varepsilon < 0$ and Na$^+$ ions repel each other, there are more vacancies available than

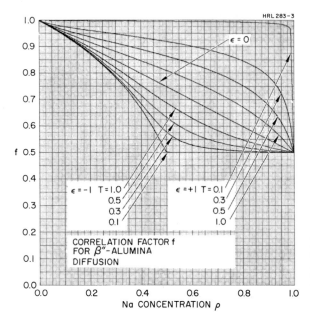

Fig. 2.11. The correlation factor f for Na* ion diffusion in β''-alumina plotted against the Na concentration ϱ [2.21]

(2.35) indicates as is seen from (2.33). For the factor W, we see from (2.34) that $W=1$ when $\varepsilon=0$. For $\varepsilon<0$, $W>1$, which indicates that Na^+ migration is enhanced because a diffusing Na^+ ion is pushed by other Na^+ ions. The effect of W can be very large when the concentration of Na^+ ions is large enough as is clear from (2.34). Therefore, a repulsive interaction among Na^+ ions (like Coulomb interaction) can enhance the diffusion greatly. The correlation factor f is plotted in Fig. 2.11. $f=1$ means that when a Na^+ ion jumps into a vacancy, the direction of its next jump is random; that means that the vacancy with which the Na^+ has exchanged sites has no direct influence on the direction of the next jump of the ion. In the region $\varrho \rightarrow 1$, f takes the value of self-diffusion irrespective of the strength of interaction. $f=0.5$ in Fig. 2.11 is this limiting value according to the pair approximation used in the present theory. The rigorous limiting value is $f=1/3$ [1]. Although the limiting value of f at $\varrho=1$ is indicative of the mechanism of diffusion as is generally accepted [2.2], f is generally a function of both the concentration of diffusing ions and the strength of interactions and, hence, indirectly of temperature.

An interesting result of the theoretical prediction is that the correction factor due to the many body effect VWf takes approximately the form of the Arrhenius equation

$$VWf = A \exp(-u'/kT). \tag{2.36a}$$

---

[1]   The discrepancy between the exact value and that for the pair approximation has now been removed [2.81].

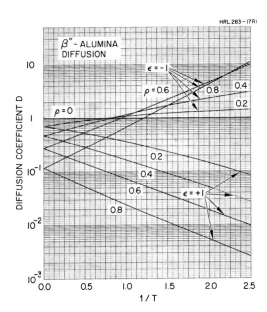

HRL 283-17RI

Fig. 2.12. Temperature dependence of $D = VWf$ for β''-alumina [2.21]

The product $D = VWf$ (note that D defined in this section is not the diffusion coefficient) is plotted as a function of the inverse temperature $1/T$ in Fig. 2.12. Temperature T here is measured in units of $|\varepsilon|/k$. The $\log D$ vs $1/T$ curves shown are almost linear. Thus (2.32) can be written as

$$D_2 = d^2 A\theta \exp[-(u+u')/kT]. \qquad (2.36b)$$

Depending on whether ε is positive or negative, u' is positive or negative and

$$u' \simeq \varepsilon. \qquad (2.37)$$

Therefore, if the interaction among diffusing ions is repulsive ($\varepsilon < 0$), the interaction eventually reduces the activation energy by u' and makes the preexponential factor smaller by the factor A. The effect is mainly due to W and hence the effect of the interaction on the preexponential factor is mainly the reduction in the attempt frequency.

The composition dependence of D is shown in Fig. 2.13. The curve for $\varepsilon = 0$ is proportional to the concentration of vacancies, $V = 1 - \varrho$, and shows that the diffusion coefficient is proportional to the number of vacancies for the random walk process. Curves for $\varepsilon < 0$ show the enhancement caused by repulsive interactions; the enhancement effect is pronounced at low temperatures and for compositions toward $\varrho \to 1$. Because the enhancement is derived from the fact that there are many more available sites than the number of diffusing ions, the result indicates that the disordered sublattice with repulsive interaction among conducting ions promotes the fast diffusion.

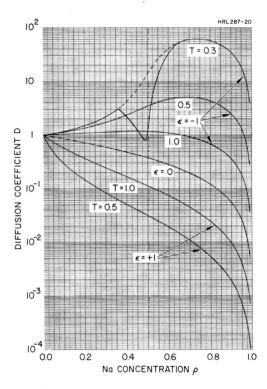

Fig. 2.13. The composition dependence of $D = VWf$ for β''-alumina [2.21]

As is pointed out in Sect. 2.1, the repulsive interaction tends to create the ordered arrangement of ions and vacancies. The long range order can eventually be established within a certain region of composition range near $\varrho = 0.5$ at low temperatures [2.21, 70]. If the long range order is established, preferred lattice sites are created among otherwise equivalent lattice sites. This increases the activation energy of motion of ions and eventually decreases the diffusion [2.32]. However, the main decrease in diffusion results from the effect of the correlation factor f. In the ordered structure, even if an ion on a preferred site jumps out into a nonpreferred site, the ion tends to go back into the original preferred site immediately. This means that $f \sim 0$. In Fig. 2.13 where the composition dependence of D is shown, a dip is observed around $\varrho = 0.5$ in the dotted curve for $T = 0.3$. The dip is due to the effect of the long range order created at this composition range at this temperature. In fact, if $\log D_2$ is plotted against $1/T$ in such a composition range where the order-disorder transition occurs, we can observe a break at the order-disorder transition point like that observed in $RbAg_4I_5$ at 209 K as mentioned earlier [2.32]. On the other hand, no dip in f is observed around $\varrho = 0.5$ in Fig. 2.11 even at low temperatures, contrary to the statement above, because f in Fig. 2.11 was calculated in the disordered state only, by suppressing the formation of long range order [2.21]. The decrease in f in the ordered state will be explained for β-alumina later.

Ionic Conductivity

An advantage of the path probability method is that the calculations of the isotope diffusion and the ionic conductivity can be made with the same degree of approximation. In formulating the ionic conductivity, we impose an external electric field E on the system and calculate the net flow of Na^+ ions migrating toward the field direction. In this case, we use $\Phi_1 \equiv Y_{+1} - Y_{-1}$ which is similar to (2.28) except that this is for the Na^+ ions rather than Φ_2 for Na^*. The ionic conductivity σ_0 is the ratio of Φ_1 to the applied field E:

$$c'\Phi_1 = \sigma_0 E \tag{2.38}$$

where c' is a normalization constant.

The effect of the applied field E comes into the theory in the form of the lowering of the activation energy toward the field direction. Again, we have two tasks. The first is to express $Y_{\pm 1}$ in terms of the most probable path and the second is to find the state variables in the stationary state. These two tasks can be carried out parallel to the case of D_2. The result shows that σ_0 contains the same V and W that appear in (2.32) but not f. Hence, we find the relation as in the random walk approach

$$\frac{D_2}{\sigma_0} = \frac{kT}{e^2 n} f \tag{2.39}$$

where e is the magnitude of the electronic charge of a Na^+ ion and n is the density of Na^+ ions per unit volume. Equation (2.39) is essentially the Nernst-Einstein relation. If the charge diffusion coefficient D_c is defined as

$$D_c = \sigma_0 kT/ne^2 , \tag{2.40}$$

the ratio $D_2/D_c = f$ is called the Haven ratio. The Haven ratio has been used to deduce the correlation factor in order to investigate the mechanism of diffusion. But since f is a function of concentration and of the stregth of interactions among ions and hence indirectly of temperature, the determination of the Haven ratio in fast ionic conductors is not of too much importance in itself.

Because of V, the ionic conductivity is proportional to $\varrho(1-\varrho)$ and not simply to ϱ. In other words, the *effective number* of the conducting ions is not $N\varrho$ where N is the number of sites per unit volume, but is $N\varrho(1-\varrho)$. In the absence of interaction, the maximum of the ionic conductivity can be expected around $\varrho = 0.5$ whereas the enhancement due to repulsive interactions shifts the maximum toward $\varrho = 1$.

2.4.3 β-Alumina

In the previous section, the case of the two dimensional honeycomb lattice with no preferential sites is treated as β″-alumina. However, in most cation disordered phases, the available sites are generally not equivalent. The case for β-alumina with two different sites is a good example for investigat ion of the role of preferential sites for ionic conduction in general. All the results shown here are for negative ε. $\varrho = 0.5$ corresponds to the stoichiometric case where the number of conducting ions is equal to the number of sites with $w = 0$ (A sites).

In Fig. 2.2, we assume that A sites are energetically more favorable than B and the difference in the energy of occupancy is assumed to be w. This extra energy w on an A sublattice induces two immediate effects. One is that more Na^+ ions tend to sit on the A sublattice sites than B in equilibrium; the second is that an additional activation energy w is needed when a Na^+ ion jumps from an A site into an adjacent B site. Therefore, the situation is somewhat similar to a built-in superlattice discussed in the previous section. Except for these two points, the basic procedures of treating the problem are the same as those presented in previous sections and the basic characteristics are nearly the same. Therefore, the results which are characteristic to such an ordered system will be discussed here.

Generally, the tracer diffusion coefficient D_2 can be written the same as (2.32). However, when we have two sublattices and one is preferred, it is more illuminating to divide each contribution into two with respect to two sublattices A and B.

The diffusion coefficient

$$D_2 = d^2 \theta \, e^{-\beta u} V W f \tag{2.41a}$$

$$\beta = 1/kT \tag{2.41b}$$

can then be divided into two

$$\frac{1}{D_2} = \frac{1}{2} \left(\frac{1}{D_A} + \frac{1}{D_B} \right) \tag{2.42}$$

where

$$D_A \equiv d^2 \theta \, e^{-\beta u} V_A W_A f \tag{2.43}$$

and

$$D_B \equiv d^2 \theta \, e^{-\beta u} V_B W_B f . \tag{2.44}$$

Here D_A and D_B mean the diffusion coefficients of tracer cations jumping from A sites and of those jumping from B sites. V_A, W_A, etc., then have corresponding meanings for the vacancy availability factor and the effective jump frequency factor. Because D_2 is a harmonic mean of D_A and D_B, D_2 is determined practically by the smaller one of D_A and D_B. Individually, V is a weighted mean of V_A and V_B

$$V = (p_{2e}/2\varrho_2)V_A + (q_{2e}/2\varrho_2)V_B = (x_{23e} + x_{32e})/2p_2 \tag{2.45a}$$

$$q_{2e} = 1 - p_{2e} \tag{2.45b}$$

while W is a harmonic mean of W_A and W_B or

$$W^{-1} = (W_A^{-1} + W_B^{-1})/2. \qquad (2.46)$$

It is natural to expect that $D_A \ll D_B$ at a highly ordered state because, for ions jumping from A sites, a pushing effect of neighboring Na^+ ions on B sites is missing and hence W_A is much smaller than W_B.

It is possible to divide D_2 in a different fashion so that f can be divided in a similar way.

$$\frac{1}{D_2'} = \frac{1}{2}\left[\left(\frac{1}{D_A'}\right) + \left(\frac{1}{D_B'}\right)\right] \qquad (2.47)$$

where

$$D_A' = d^2 \theta e^{-\beta u} V W f_{AB} \qquad (2.48)$$

$$D_B' = d^2 \theta e^{-\beta u} V W f_{BA} \qquad (2.49)$$

and

$$f^{-1} = (1/2)(f_{AB}^{-1} + f_{BA}^{-1}). \qquad (2.50)$$

f_{AB} has a meaning of a partial correlation factor for cations jumping from A sites, whereas f_{BA} is that for cations jumping from B sites. When a Na^+ ion jumps from A to B, the probability of its going back to the original A sites is high so that $f_{AB} \rightarrow 0$ for a highly ordered state. Because f is determined by the smaller of f_{AB} and f_{BA}, f becomes very small for the ordered states. The concentration dependence of f at several temperatures is plotted in Fig. 2.14. As expected, there is a sharp dip of f at $\varrho = 0.5$ at low temperatures. Here, the calculation is made for $|\varepsilon| = |w|$ and the temperature is again measured in terms of $|\varepsilon|/k$. The behavior is also similar for $w = 0$ (β''-alumina) when the long range order of cation distribution sets in as a result of mutual repulsive interactions among Na ions.

Calculated results for $D = VWf$ for β-alumina are shown in Figs. 2.15 and 2.16. Again D is approximately represented by an Arrhenius-type equation with a negative activation energy except at extremely high temperatures. The curvature at high temperatures depends on the magnitude of w. The sharp dip of D in Fig. 2.16 at $\varrho = 0.5$ reflects the sharp dip of f in Fig. 2.14. In other words, the decrease in the diffusion coefficient or the ionic conductivity in the ordered state is mainly due to the decrease in the correlation factor. The dip in D_2 at $\varrho = 0.5$ at low temperatures also means that the diffusion coefficient for stoichiometric solids is small unless the ions are excited to interstitial sites. The diffusion at ϱ slightly larger than 0.5, on the other hand, corresponds physically to the diffusion with the interstitialcy mechanism.

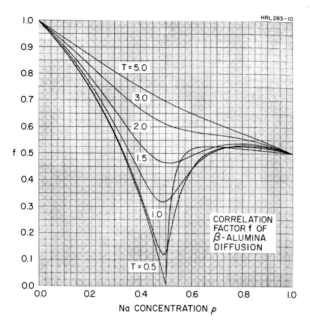

Fig. 2.14. The correlation factor f for Na* ion in β-alumina [2.21]

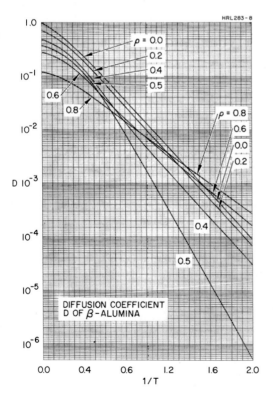

Fig. 2.15. Temperature dependence of $D = VWf$ for β-alumina [2.21]

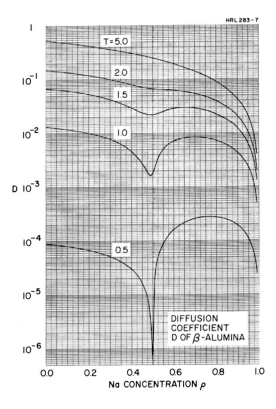

Fig. 2.16. The concentration dependence of $D = VWf$ for β-alumina [2.21]

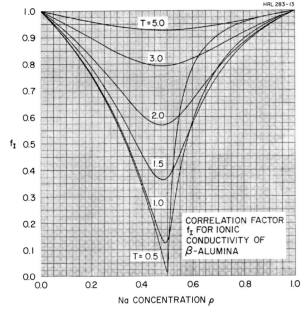

Fig. 2.17. The correlation factor f_I for ionic conductivity of Na ions in β-alumina [2.21]

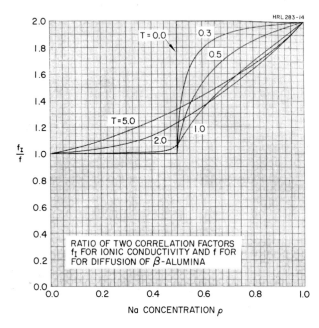

Fig. 2.18. The concentration dependence of the inverse of the Haven ratio for β-alumina [2.21]

The ionic conductivity σ_0 is defined the same way as in (2.38). Different from (2.39), however, the case of β-alumina has the relation:

$$\frac{D_2}{\sigma_0} \equiv \frac{kT}{e^2 n} \frac{f}{f_1}. \tag{2.51}$$

The factor f_1 is the correlation factor which appears in the ionic conductivity and this is the result of the physical correlation effect among conducting ions which appears in the ordered state [2.82]. The curve f_1 is shown in Fig. 2.17. Similar to f in Fig. 2.14, the dip of f_1 at $\varrho = 0.5$ is noteworthy. This confirms the statement that the decrease in the ionic conductivity in the ordered state is primarily due to the decrease in the correlation factor and not to the increase in the activation energy.

f/f_1 is the Haven ratio and is plotted in Fig. 2.18. By taking the ratio, the physical correlation effect due to the ordering of Na^+ ions is cancelled out and the ratio behaves more smoothly. Although it is a common practice to obtain the correlation factor f from the Haven ratio, the value obtained actually is f/f_1 and not f. This is conceptually quite important when the physical correlation effect exists as in the ordered case.

In principle, these two idealized cases should represent general qualitative characteristics of fast ionic conduction. The results also include some features of cooperative motion of ions like the interstitialcy mechanism. The Ising model formalism taken is suitable for incorporating different features of ionic motion as well as complicated crystallographic features [2.76] and corresponds to the extension of the random walk approach. In view of these advantages, this method is expected to be highly useful to elucidate many features of ionic conduction of a variety of practical materials. On the other hand, the assumption of the Markovian process in the present treatment can lead only to the description of the dc conductivity. It is highly desirable, in this respect, to extend the present method so that quantitative descriptions of $\sigma(\omega)$ can be obtained.

Acknowledgement. The author is indebted to valuable discussions with K. Gschwend, Y. Hirotsu, R. Kikuchi, G. L. Liedl, and R. W. Vest. The help of K. Gschwend in the preparation of the manuscript is most gratefully acknowledged. Work in this area at Purdue University has been supported by the National Science Foundation through both individual grants and the MRL program.

References

2.1   A.B.Lidiard: Ionic Conductivity. In *Handbuch der Physik*, Vol. 20, ed. by S.Flugge, (1957) p. 246
2.2   J.R.Manning: *Diffusion Kinetics for Atoms in Crystals* (Van Nostrand, Princeton, New Jersey 1968)
2.3   C.P.Flynn: *Point Defects and Diffusion* (Clarendon Press, Oxford 1972)
2.4   L.W.Strock: Z. Physik. Chem. B25, 411 (1934);—B31, 132 (1936)
2.5   C.A.Beevers, M.A.S.Ross: Z. Krist. 77, 255 (1931)
2.6   C.R.Peters, M.Bettman, J.Moor, M.Glick: Acta Cryst. B27, 1826 (1971)
2.7   W.L.Roth, R.Reidinger, S.LaPlaca: Studies of Stabilization and Transport Mechanism in Beta and Beta Alumina by Neutron Diffraction. In *Superionic Conductors*, ed. by G.D. Mahan, W.L. Roth (Plenum Press New York 1976) p. 223
2.8   H.Wiedersich, S.Geller: Properties of Highly Conducting Halide and Chalcogenide Solid Electrolytes. In *The Chemistry of Extended Defects in Non-Metallic Solids*, ed. by L.Eyring, M.O'Keeffe (North-Holland, Amsterdam 1970) p. 629
2.9   W.vanGool: Fast Ionic Conduction. In *Annual Review of Materials Science*, 4, 311 (1974)
2.10 M.O'Keeffe, B.G.Hyde: Phil. Mag. 33, 219 (1976)
2.11 C.Tubandt, F.Lorenz: Z. Phys. Chem. 87, 543 (1913). See also A.Kvist, A.-M.Josefson: Z. Naturforsch. 23a, 625 (1968)
2.12 M.O'Keeffe: Phase Transitions and Translational Freedom in Solid Electrolytes. In *Superionic Conductors*, ed. by G.D.Mahan, W.L.Roth (Plenum Press, New York 1976) p. 101
2.13 C.E.Derrington, A.Narrotsky, M.O'Keeffe: Solid State Commun. 18, 47 (1976)
2.14 T.Takahashi: J. Appl. Electrochem. 3, 79 (1973)
2.15 A.Kvist, A.Lunden: Z. Naturforsch. 219, 1509 (1966)
2.16 H.H.Möbius: Z. Chem. 2, 100 (1962)
2.17 A.R.West: J. Appl. Electrochem. 3, 327 (1973)
2.18 C.E.Derrington, A.Lindner, M.O'Keeffe: J. Solid State Chem. 15, 171 (1975)
2.19 C.E.Derrington, M.O'Keeffe: Solid State Commun. 15, 1175 (1974)
2.20 L.E.Nagel, M.O'Keeffe: Highly-Conducting Fluorides Related to Fluorite and Tysonite. In *Fast Ion Transport in Solids*, ed. by W.vanGool (North Holland, Amsterdam 1973) p. 165
2.21 H.Sato, R.Kikuchi: J. Chem. Phys. 55, 677 (1971)
2.22 R.Kikuchi: Phys. Rev. 81, 988 (1951)
2.23 H.Sato: Order-Disorder Transformation. In *Physical Chemistry*, Vol. 10, ed. by W.Jost (Academic Press, New York 1970) p. 579
2.24 B.A.Huberman: Phys. Rev. Letters 32, 1000 (1974)
2.25 M.J.Rice, S.Strässler, G.A.Toombs: Phys. Rev. Letters 32, 596 (1974)
2.26 D.O.Welch, G.J.Dienes: Bull. Am. Phys. Soc. II. 21, 286 (1976)
2.27 B.B.Owens: Solid Electrolyte Batteries. In *Advances in Electrochemistry and Electrochemical Engineering*, Vol. 8, ed. by C.W.Tobias (John Wiley and Sons, Inc., New York 1971) p. 1

2.28 W.V.Johnston, H. Wiedersich, G.W.Lindberg: J. Chem. Phys. **51**, 3739 (1969)
2.29 F.Lederman, M.B.Salamon: Bull. Amer. Phys. Soc. **20**, 331 (Mar. 1975)
2.30 B.B.Owens, G.R.Argue: Science **157**, 308 (1967)
2.31 S.Geller: Phys. Rev. B**14**, 4345 (1976)
2.32 R.Kikuchi, H.Sato: J. Chem. Phys. **51**, 161 (1969); **57**, 4962 (1972)
2.33 W.J.Pardee, G.D.Mahan: J. Solid State Chem. **15**, 310 (1975); G.D.Mahan: Phys. Rev. B**14**, 780 (1976)
2.34 S.Geller: Science **157**, 310 (1967)
2.35 M.S.Whittingham: Science **192**, 1126 (1976)
2.36 M.S.Whittingham, F.R.Gamble,Jr.: Mat. Res. Bull. **10**, 363 (1975)
2.37 C.Fouassier, C.Delmas, P.Hagenmuller: Mat. Res. Bull. **10**, 443 (1975)
2.38 C.Delmas, C.Fouassier, P.Hagenmuller: C. R. Acad. Sci. Paris **281**C, 587 (1975)
2.39 C.T.Chudley, R.J.Elliott: Proc. Phys. Soc. **77**, 353 (1961)
2.40 W.Buhrer, W.Hälg: Helv. Phys. Acta **47**, 27 (1974)
2.41 A.F.Wright, D.E.F.Fender: J. Phys. C (in press)
2.42 R.M.Cava, B.J.Wuensch: The Distribution of Silver Ions in α-AgI. In *Superionic Conductors*, ed. by G.D.Mahan, R.L.Roth (Plenum Press, New York and London 1976) p. 217
2.43 K.Funke, J.Jost: Adad, Wiss. Göttingen, Nr. **15**, 137 (1969)
2.44 R.C.Hanson, T.A.Fjeldly, H.D.Hochheimer: Phys. Stat. Sol. (b) **70**, 567 (1975)
2.45 K.Funke, J.Kalus, R.E.Lechner: Solid State Commun. **14**, 1021 (1974)
2.46 G.Eckold, K.Funke, J.Kalus, R.E.Lechner: Phys. Lett. **55**A, 125 (1975)
2.47 G.Eckold, K.Funke, J.Kalus, R.E.Lechner: J. Phys. Chem. Solids (to be published)
2.48 P.A.Egelstaff: *An Introduction to the Liquid State* (Academic Press, New York 1967)
2.49 B.A.Huberman, P.N.Sen: Phys. Rev. Letters **33**, 1379 (1974)
2.50 P.Bruesch, S.Strässler, H.R.Zeller: Phys. Stat. Sol. (a) **31**, 217 (1975)
2.51 P.Fulde, L.Pietronero, W.R.Schneider, A.Strassler: Phys. Rev. Letters **26**, 1776 (1971)
2.52 H.R.Zeller, P.Bruesch, L.Pietronero, S.Strässler: Lattice Dynamics and Ionic Motion in Superionic Conductors. In *Superionic Conductors*, ed. by G.D.Mahan, W.L.Roth (Plenum Press, New York 1976) p. 201
2.53 Y.F.Y.Yao, J.T.Kummer: J. Inorg. Nucl. Chem. **89**, 2453 (1967)
2.54 N.A.Toropov, M.M.Stukalova: Cr. Acad. Sci. U.R.SS **27**, 974 (1943);—Cr. Acad. Sci. U.R.SS **24**, 459 (1939)
2.55 J.T.Kummer: β-Alumina Electrolytes. In *Prog. Solid. State Chemistry*, Vol. 4, ed. by J.O.McCaldin (Pergamon Press, Oxford, 1972) p. 60
2.56 S.J.Allen,Jr., J.P.Remeika: Phys. Rev. Letters **33**, 1478 (1974)
2.57 J.C.Wang, M.Gaffari, Sang-Il Choi: J. Chem. Phys. **63**, 772 (1975)
2.58 W.H.Flygare, R.A.Huggins: J. Phys. Chem. Solids **34**, 1199 (1973)
2.59 R.A.Huggins: Very Rapid Ionic Transport in Solid. In *Diffusion in Solids*, ed. by A.S.Nowick, J.J.Burton (Academic Press, New York 1975) p. 445
2.60 M.Born, K.Huang: *Dynamical Theory of Crystal Lattices* (Oxford University Press 1954)
2.61 L.Pauling: Z. Krist. **67**, 377 (1928)
2.62 H.Reik: The small Polaron Concept and its use for the interpretation of *DC* and Optical Properties of Semiconducting Transitional Metal Oxides. In *Polarons in Ionic Crystals and Polar Semiconductors*, ed. by J.T.Devreese (North-Holland, London 1972) p. 679
2.63 C.P.Flynn, A.M.Stoneham: Phys. Rev. B**1**, 3906 (1970)
2.64 G.D.Mahan, W.J.Pardee: Phys. Lett. **49**A, 325 (1974)
2.65 I.Yokota: J. Phys. Soc. Japan **21**, 420 (1966)
2.66 W.vanGool: J. Solid State Chem. **7**, 55 (1973)
2.67 W.vanGool, P.H.Bottelberghs: J. Solid State Chem. **7**, 59 (1973)
2.68 W.L.Roth: J. Solid State Chem. **4**, 60 (1972)
2.69 C.Kittel, J.K.Galt: Ferromagnetic Domain Theory. In *Solid State Physics*, Vol. 3, ed. by F.Seitz, D.Turnbull (Academic Press, New York 1956) p. 439
2.70 K.R.Subbadwarrey, G.D.Mahan: Phys. Rev. Letters **37**, 642 (1976)
2.71 G.D.Mahan: Theoretical Issues in Superionic Conductors. In *Superionic Conductors*, ed. by G.D.Mahan, W.L.Roth (Plenum Press, New York 1976) p. 115

2.72 W. van Gool: Domain Model for Superionic Conductors. In *Superionic Conductors*, ed. by G.D.Mahan, W.L.Roth (Plenum Press, New York 1976) p. 143

2.73 M.J.Rice, W.L.Roth: J. Solid State Chem. **4**, 294 (1972)

2.74 G.W.Haas: J. Solid State Chem. **7**, 155 (1973)

2.75 R.Kikuchi, H.Sato: J. Chem. Phys. **55**, 702 (1971)

2.76 R.Kikuchi: Cation Diffusion and Conductivity in Solid Electrolytes. In *Fast Ion Transport in Solids*, ed. by W. van Gool (North-Holland/American Elsevier, Amsterdam 1973) p. 249

2.77 H.Sato, R.Kikuchi: Path Probability Method as Applied to Problems of Superionic Conduction. In *Superionic Conductors*, ed. by G.D.Mahan, W.L.Roth (Plenum Press, New York 1976) p. 135

2.78 D.M.Burky: Closed Form Approximations for Lattice Systems. In *Phase Transitions and Critical Phenomena*, Vol. 2, ed. by C.Domb, M.S.Green (Academic Press, New York 1972) p. 329

2.79 R.Kikuchi: Progr. Theoret. Phys. (Kyoto) Suppl. **35**, 1 (1966)

2.80 N.L.Peterson: Diffusion in Metals. In *Solid State Physics*, Vol. 22, ed. by F.Seitz, D.Turnbull, H.Ehrenreich (Academic Press, New York 1968) p. 409

2.81 K.Gschwend, H.Sato, R.Kikuchi: Bull. Am. Phys. Soc. **22**, 442 (1977)

2.82 H.Sato, R.Kikuchi: Correlation Factor and Nernst-Einstein Relation in Solid Electrolytes. In *Proc. 11th University Conf. on Ceramics*, ed. by A.R.Cooper, A.Heuer (Plenum Press, New York 1975) p. 1149

3. Halogenide Solid Electrolytes

S. Geller

With 8 Figures

In this chapter, we emphasize the structural features which are the cause of the high conductivities of the halogenide solid electrolytes. Results of various relevant measurements are discussed. A survey of the many halogenide compounds found to be solid electrolytes is included.

3.1 AgI-Based Solid Electrolytes

3.1.1 Crystal Structures

Rubidium Silver Iodide, RbAg$_4$I$_5$

There are three crystalline modifications of RbAg$_4$I$_5$, labelled α, β, γ in order of decreasing temperature. Crystals of α-RbAg$_4$I$_5$ belong to space group P4$_1$32(O^7) or P4$_3$32(O^6) [3.1] (both have been made [3.2]) with lattice constant $a = 11.24$ Å, and the unit cell containing 4 RbAg$_4$I$_5$. The 20 iodide ions are arranged such that they form tetrahedra which share faces (Fig. 3.1) thereby forming a network of pathways through which the Ag$^+$ ions can move by essentially a hopping mechanism (see Chap. 2). The large Rb$^+$ ions are surrounded by highly distorted iodide octahedra and are not mobile. α-RbAg$_4$I$_5$ has the highest room temperature specific conductivity, 0.27 [Ω cm]$^{-1}$ [3.3–5], of any solid ionic material, and this implies that it thus far has the most effective solid electrolyte structure at room temperature.

In each unit cell of α-RbAg$_4$I$_5$, there are 56 sites for the 16 Ag$^+$ ions: one 8-fold set designated Ag(c) and two 24-fold sets designated Ag(II) and Ag(III). The latter are general positions of the space group. The sites belonging to the general sets alternate in channels which are perpendicular to each unit cell face (see Fig. 3.1). There are two such channels per face. Because the channels are formed by face-sharing tetrahedra, Ag$^+$ ions moving through them must zig-zag between sites. Each channel increment within the unit cell contains eight sites, four each of Ag(II) and Ag(III).

The channels are joined by Ag(II) as well as by Ag(c) sites, but Ag(c) sites are not actually required for diffusion of Ag$^+$ ions through the crystal. Only half the Ag(II) sites can be occupied at any instant because the other half are too close to be occupied simultaneously. Because of nearest neighbor constraints, if half the Ag(II) sites were occupied, there could be at *most* 12 Ag(III) and 4 Ag(c) sites available to the remaining Ag$^+$ ions.

Fig. 3.1. Model of the structure of $RbAg_4I_5$. Large spheres represent iodide ions, smaller white spheres, Rb^+ ions. Centers of sleeves on short horizontal arms designate possible positions of Ag^+ ions. Note the channels at the upper middle and lower left. (From [3.1]; copyright 1967 by the American Association for the Advancement of Science)

The three types of sites are crystallographically nonequivalent and the second nearest neighbor surroundings of each type are quite different. Thus it should be expected that their site energies and, therefore, their occupancies would be different. At room temperature, they are respectively: 0.9 ± 0.3, 9.4 ± 0.9, 5.5 ± 0.8; these figures give the *limits of error*, 3σ. (These large limits of error result from interactions between the multipliers and thermal parameters.) In the final refinement of the crystal structure, no constraint was put on the total Ag^+ ion unit cell content. The convergence to the correct total gives assurance that the limits of error are conservative.

The solid electrolyte $RbAg_4I_5$ was discovered by *Bradley* and *Greene* [3.6] and independently by *Owens* and *Argue* [3.7]. KAg_4I_5 and $NH_4Ag_4I_5$ [3.7, 8] and the solid solutions [3.7] $(K_{0.5}Rb_{0.5})Ag_4I_5$, $(K_{0.75}Rb_{0.25})Ag_4I_5$ and $(K_{0.5}Cs_{0.5})Ag_4I_5$ are solid electrolytes isostructural with $RbAg_4I_5$. A compound KCu_4I_5, stable between 257 and 332 °C [3.7], shown [3.9] by powder x-ray diffraction to be isostructural with $RbAg_4I_5$, is also a solid electrolyte [3.10]. About 0.06 Ag^+ ion in $RbAg_4I_5$ can be replaced by Cu^+ when copper electrodes are used in electrolysis experiments [3.11]. At least 20 atomic percent of Ag^+ in $RbAg_4I_5$ can be replaced by Cu^+ at 150° C as determined by x-ray powder diffraction photography [3.12].

The $\alpha - \beta$ transition of $RbAg_4I_5$ was first observed by *Geller* [3.1] in two ways: by noticing the onset of birefringence of a specimen as it was cooled through the transition and by means of x-ray powder diffractometry. The precise transition temperature, 209 K, was obtained by *Johnston* et al. [3.13] from heat capacity measurements; *Lederman* et al. [3.14] report 208 K for this transition. The $\alpha - \beta$ transition is second order: this was first surmised from the early [3.1] optical and x-ray diffraction examinations, from the conductivity vs temperature measurements, which, at first [3.7] did not seem to show it at all (but see [3.3]), by the heat capacity measurements [3.13, 14] and finally by a detailed "single-crystal" structure analysis [3.15].

Table 3.1. Space groups and equipoint transformations of $RbAg_4I_5$ phases (numbers in parentheses designate point symmetries)

$T > 209$ K $P4_132$	209 K $> T > 122$ K R32	$T < 122$ K P321
Rb in 4a (32)	$\left\{\begin{array}{l} 1a(32) \\ \\ 3e(2) \end{array}\right.$	$\left\{\begin{array}{l} 1a(32) \\ 2d(3) \end{array}\right.$ $\left\{\begin{array}{l} 3f(2) \\ 6g(1) \end{array}\right.$
I, Agc in 8c (3)	$\left\{\begin{array}{l} 2c(3) \\ \\ 6f(1) \end{array}\right.$	$\left\{\begin{array}{l} 2c(3) \\ 2 \text{ sets } 2d(3) \end{array}\right.$ 3 sets 6g(1)
I in 12d (2)	$\left\{\begin{array}{l} 3d(2) \\ \\ 3e(2) \\ \\ 6f(1) \end{array}\right.$	$\left\{\begin{array}{l} 3e(2) \\ 6g(1) \end{array}\right.$ $\left\{\begin{array}{l} 3f(2) \\ 6g(1) \end{array}\right.$ 3 sets 6g(1)
AgII, AgIII in 24e	$\left\{\begin{array}{l} 4 \text{ sets of } 6f \\ \text{each; 8 sets} \\ \text{of 6f, total} \end{array}\right.$	$\left\{\begin{array}{l} \text{'3 sets 6g each.} \\ 24 \text{ sets of } 6g, \\ \text{total} \end{array}\right.$

The quotation marks are used around "single-crystal" because the crystal from which the x-ray intensity data were collected was not *strictly* single even though it was so at room temperature. That is to say, when a single crystal of α-$RbAg_4I_5$ is cooled through the transition temperature, it becomes multiply twinned. To simplify absorption correction, the x-ray intensity data are obtained from a very small spherically shaped crystal; thus there is no obvious way of avoiding this twinning. If at room temperature, the crystal was originally aligned along a fourfold axis, then below the transition this axis is a pseudo-fourfold axis. This implies that the overall twin volumes are equal and that the distortion from the cubic structure is small, which was already seen in the x-ray powder diffraction pattern [3.1] from which it was deduced that the transition is to a rhombohedral structure.

Because the α−β transition is second order, the space group to which β-$RbAg_4I_5$ belongs must be a subgroup of $P4_132$ or $P4_332$, most probably $R32(D_3^7)$. At 130 K, the lattice constants are $a = 11.17 \pm 0.01$ Å, $\alpha = 90.1 \pm 0.05°$. Table 3.1 gives the equipoint transformations involved in the phase change. The displacements of ions and equilibrium Ag^+ ion sites, relative to the α-phase, are small, the largest, 0.3 Å, being those for two Ag^+ ion sites. The largest displacement of an I^- ion is 0.1 Å. The most significant difference between the α- and β-phases is the *preferential* distribution of Ag^+ ions, in the latter, over sets of sites formerly crystallographically equivalent. The α−β transition is designated a disorder-disorder transition because neither the α- nor the β-phase can actually be ordered.

The $\beta - \gamma$ transition was first observed in conductivity vs temperature measurements [3.7] and then by x-ray diffraction [3.1]. Heat capacity measurements [3.13] confirmed that the transition is first order and occurs at 122 K. Measurements of the dependence of elastic constants on pressure and temperature [3.16] with ultrasonic waves cause the crystal to fracture when it reaches the $\beta - \gamma$ transition. However the crystal is not destroyed in cooling and warming cycles alone, because the γ-phase is still closely related to the β-phase of which it is a "superstructure" [3.15].

The γ-RbAg$_4$I$_5$ phase [3.15] is trigonal and belongs to space group P321(D_3^2) with $a = 15.776 \pm 0.005$ Å, $c = 19.320 \pm 0.005$ Å at 90 K. The dimensions are very nearly cubic: $11.155\sqrt{2} = 15.776$, $11.155\sqrt{3} = 19.321$. The unit cell is three times as large as the cubic unit cell would be at 90 K and therefore contains 12 RbAg$_4$I$_5$. The equipoint transformations from β-RbAg$_4$I$_5$ to γ-RbAg$_4$I$_5$ are given in Table 3.1. The structure of the γ-phase can be derived from the β-phase by first transforming the latter to a triply primitive hexagonal cell and then permitting further displacements of ions and Ag$^+$ ion sites with respect to their positions in the β-phase, thus resulting in a primitive cell. Under the assumption that 30 Ag$^+$ ions fill (transformed) Ag(II)-type sites, there are 146 possible ordering schemes in the trigonal crystal. However, there is evidence that the γ-phase is not ordered at 90 K (even though unlike the cases of the α- and β-phases, it could be). The single most probable arrangement for such an ordered phase has been deduced [3.15]. It is probable, as predicted earlier [3.13, 17], that at least over part of the temperature range below 122 K, the $\beta - \gamma$ transition temperature, the phase is still a solid electrolyte.

Unfortunately, the detailed information on the β- and γ-RbAg$_4$I$_5$ phases was not available at the time that *Pardee* and *Mahan* published their paper [3.18] on these transitions. As pointed out in [3.15], when the symmetry of a crystal decreases, the site-degeneracy decreases; that is, the number of sets of crystallographically equivalent sites increases. In the RbAg$_4$I$_5$ case, there are three such sets in the α-phase which go into ten sets in the β-phase. Of these, three are essentially empty at 130 K and none of the others is filled; in fact, most are *constrained* to be only partially filled.

If this is considered to be order, then no theory is needed, because it should be clear that *some* change in site preference is implied by a transition to lower symmetry. As indicated above, β-RbAg$_4$I$_5$ cannot have an ordered arrangement of the Ag$^+$ ions, and one can ascertain this only from a knowledge of its detailed structure.

The $\beta - \gamma$ transition is not necessarily a *Pardee-Mahan* Class I transition either, because it is not an electrolyte-insulator transition. This possibility *was* pointed out earlier [3.13, 17] and has since been given further credence [3.15].

Tetramethylammonium Silver Iodide [(CH$_3$)$_4$N]$_2$Ag$_{13}$I$_{15}$

[(CH$_3$)$_4$N]$_2$Ag$_{13}$I$_{15}$ is one of many tetraalkylammonium iodide solid electrolytes discovered by *Owens* [3.3, 19, 20]. It has one of the higher room temperature conductivities in this group and was therefore one of the early crystals of which a detailed structure determination was made. The structure [3.21] of $[(CH_3)_4N]_2Ag_{13}I_{15}$ belongs to space group R 32 (D_3^7) and contains one formula unit in the rhombohedral cell with dimensions $a = 11.52 \pm 0.02$Å, $\alpha = 67.35 \pm 0.15°$. The triply primitive hexagonal cell has dimensions $a = 12.77 \pm 0.03$, $c = 26.54 \pm 0.05$ Å. As in the case of $RbAg_4I_5$, all the iodide ions are at the corners of tetrahedra which share faces, but the structure of $[(CH_3)_4N]_2Ag_{13}I_{15}$ is substantially more complicated than that of $RbAg_4I_5$, accounting, in part, for its lower conductivity.

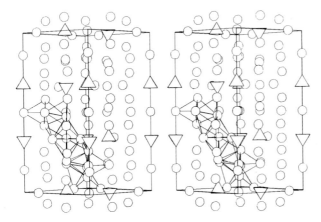

Fig. 3.2 Stereoscopic pair of unit cells of $[(CH_3)_4N]_2Ag_{13}I_{15}$. One of each type of channel is illustrated by connections between iodide ions. Ag^+ ion positions are not shown. (From [3.21])

The tetrahedra are arranged in two kinds of channels (Fig. 3.2), one of which ends at the crystal surfaces. The other, consisting of ten tetrahedra each, begins and ends at a tetramethylammonium ion. The channels are themselves interconnected. There are eight crystallographically nonequivalent sets of sites for the Ag^+ ions. The sites in one of these sets are not directly in channels, but do link channels. Thus, as in the $RbAg_4I_5$ case, the network is three dimensional. There are 123 tetrahedra and 39 Ag^+ ions in the triply primitive unit cell. The distribution of the Ag^+ ions over the eight sets of crystallographically nonequivalent sites is significantly nonuniform; in fact, one set of sites is (within experimental error) empty, meaning that its energy is quite high, or equivalently that the "lifetime" of a Ag^+ ion in such a site is too short to be "seen" by the x-ray diffraction analysis.

The Ag^+ ion sites in the triply primitive cell are in six general 18-fold sets and two special sets, one 6-fold and one 9-fold, of the space group. Inasmuch as there are 39 Ag^+ ions in the triply primitive unit cell, there is absolutely no way in which the Ag^+ ions in this crystal can be ordered. No evidence of a transition is seen in the conductivity vs temperature plot down to 220 K [3.3, 19]; there may be one at a still lower temperature. However, it should be emphasized that there is no *requirement* that a transition from an insulator to a solid electrolyte exist.

The Pyridinium Iodide-Silver Iodide System

For convenience we shall denote the pyridinium ion, $(C_5H_5NH)^+$, by Py^+. There are two compounds in the PyI-AgI system that are solid electrolytes, namely $PyAg_5I_6$ and $Py_5Ag_{18}I_{23}$.

$PyAg_5I_6$

At room temperature, $PyAg_5I_6$ belongs to space group $P\,6/mcc$ (D_{6h}^2). The hexagonal cell has lattice constants $a = 12.03 \pm 0.02$, $c = 7.43 \pm 0.01$ Å and contains two $(C_5H_5NH)Ag_5I_6$ [3.22]. At 240 K, the Ag^+ ions appear to be ordered within experimental error [3.23]. *Hibma* [3.24] has found a second order transition to a monoclinic structure at ~ 230 K and more recent work by *Hibma* and *Geller* [3.25] indicates one more transition at approximately 180 K also to a monoclinic structure with a different orientation of the monoclinic axes. The last appears to be a first order transition, even though it is not *clearly* indicated by the specific heat data. These low temperature transitions must involve the ordering of the pyridinium ions. It is noteworthy that in PyI itself, the pyridinium ions are disordered [3.26] and that *Hibma* [3.24] has found a clearly first order transition of PyI at 250 K with specific heat measurements.

A plan view of the crystal structure of $PyAg_5I_6$ at $-30\,°C$ is shown in Fig. 3.3. It seems at first view to be a simple structure relative even to that of $RbAg_4I_5$. The iodide ions are in one set of positions designated 12l in the International Tables for x-ray Crystallography (Ref. [3.27], p. 300). They are arranged in layers at $z = 0$ and $1/2$ in the unit cell. The pyridinium ions, which are in *partial* static rotational disorder, are centered on the c-axis at $z = \pm 1/4$. They are surrounded by 12 iodide ions, and although it *seems* in Fig. 3.3 that there is much empty space between the Py^+ ions and the surrounding I^- ions, the H atoms, not shown, fill this space quite efficiently. At $-30\,°C$, the ten Ag^+ ions per unit cell fill two sets of sites, designated 6f and 4c in the "International Tables". These are also in layers $z = \pm 1/4$. Another set of sites, which is a general set, designated 24m, is empty at $-30\,°C$. Half of these per unit cell cen be seen by careful examination of Fig. 3.3. (To see the other half requires the addition of another layer of I^- ions.) *Every* iodide ion is at the corner of one of these "m-type" tetrahedra.

An important feature of the structure of $PyAg_5I_6$ is that it contains four iodide octahedra per unit cell, the Ag^+ ion sites at the center of these being the 4f positions. These octahedra share faces that are parallel to the hexagonal (001) plane. Thus there are *straight* channels along the c-axis in this structure. The remaining six faces of the octahedra are shared with the m-type tetrahedra. The m-type tetrahedron shares one face with an octahedron, one face with an f-type tetrahedron, one face (parallel to 001) with another m-type tetrahedron and the fourth face opens to the Py^+ ion and is therefore not involved in the diffusion of Ag^+ ions in the crystal. The f-type tetrahedra share four faces with m-type tetrahedra but only two *edges* with two other f-type tetrahedra (see Fig. 3.3).

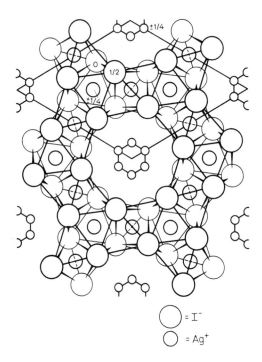

$\bigcirc = I^-$

$\bigcirc = Ag^+$

Fig. 3.3. Plan view of structure of $PyAg_5I_6$ at $-30\,°C$, at which temperature the Ag^+ ions are ordered. (From [3.22])

The octahedral (c) sites are sufficiently far from each other, 3.71 Å, so that no problem arises in the complete filling of the octahedral sites. Further, it is possible for Ag^+ ions to occupy both m- and c-sites; the intersite distance is 2.7 Å. It is not possible for an f-site and an immediately neighboring m-site to be simultaneously occupied because their intersite distance is 1.60 Å. Nor is it possible for an m-site and an immediately neighboring m-site to be simultaneously occupied because their intersite distance is 2.0 Å.

As the temperature is increased from $-30\,°C$, Ag^+ ions in the c- and f-sites are excited into the m-sites (see Fig. 3.3). The process cannot *begin* with Ag^+ ions moving from the c- to the m-sites for reasons given above, but there is no problem for the process to begin with Ag^+ ions moving from the f-sites to the m-sites. In fact, when an ion in an f-site moves to an m-site, it frees three other m-type tetrahedra for occupancy by Ag^+ ions from the octahedra.

The channels parallel to the c-axis, formed by the face-sharing octahedra, are the simplest observed in any solid electrolyte. Their importance will become more evident in a later discussion of conductivity. The other pathways for the Ag^+ ions are considerably more complicated. In *approximately* the (001) plane, they are circuitous, about the Py^+ ions, e.g., $f \to m \to c \to m \to f \to$, etc. These, of course, need not occur in a circle about a single Py^+ ion, but can, if an electric field is applied perpendicular to the c-axis, move resultantly in the direction of the field. Similarly it is possible for the Ag^+ ions to move in directions skew to the c-axis. Thus, $PyAg_5I_6$ is a three-dimensional solid electrolyte.

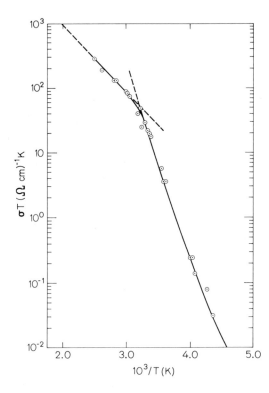

Fig. 3.4. Specific conductivity multiplied by absolute temperature vs reciprocal of absolute temperature for $PyAg_5I_6$. (From [3.22])

Shown in Fig. 3.4 is a plot of $\log(\sigma \cdot T)$ vs $1/T$ for polycrystalline $PyAg_5I_6$. This figure shows a transition at ~ 325 K. The transition was also observed by use of a differential calorimeter and has been confirmed by *Hibma* [3.24] with differential calorimetric measurements and with conductivity measurements on single crystals.

The interpretation suggested by *Geller* and *Owens* [3.22], who first reported this phenomenon, is that the transition is of a disorder-disorder type, i.e., from lower to higher disorder. In the lower temperature phase, not all the Ag^+ ions are completely mobile, while in the high temperature phases, all are mobile. The discussion above shows that this is possible in this structure. However, it should be emphasized that at temperatures above the transition, the distribution of Ag^+ ions over the three crystallographically nonequivalent sets of sites is always significantly nonuniform as should be expected.

Pardee and *Mahan* [3.18] suggest that the phase transition of $PyAg_5I_6$ at ~ 325 K is of the order-disorder type. In this case, the Ag^+ ions can and do order at approximately 240 K, so that their thesis does have a basis. On the other hand, more recent results of *Hibma* [3.24] indicate that there is neither a first nor second order phase change at 325 K because of the broadness of the peak in the specific heat vs temperature curve. However, there is evidence [3.22] from the I^- ion positional parameters that some sort of phase change does take place.

Landau and *Lifschitz* (Ref. [3.28], Chap. XIV) indicate that in a second order transition, the space group of one phase must be a subgroup of the other. Based on this criterion alone, we cannot call the phase change at 325 K second order. Nor do the collective data permit us to call it first order. However, from an atomistic point of view, there seem to be three phases belonging to the same space group. Whether or not these phases can be said strictly to be related by second order transitions is at this point moot. At 240 K, the Ag^+ ions are ordered; this is one phase. When the Ag^+ ions move into the m-type sites, we really have a different crystal structure: sites which were formerly empty now are partially occupied, and sites that were previously full are no longer so. The *description* of the structure requires the inclusion of new positions. Now the physical properties of a crystal also must play a role in determining its structural nature (as well as the converse), and we know that not all the Ag^+ ions are mobile below the 325 K transition. That is to say, between 240 and 325 K, there are certain constraints on the Ag^+ ions which are not present at temperatures greater than 325 K. The constraints *are* structural as indicated earlier. Thus (except for the Py^+ ions) there is an ordered phase between about 220 and 240 K, a phase of "low" disorder between about 240 and 325 K, and a phase of "high" disorder above 325 K. For the time being, we refer to these phases as β', β, and α, respectively.

It is of particular interest to be able to find an independent method of determining the number of carriers in the two high temperature phases of $PyAg_5I_6$. It does not seem possible yet to interpret the results of Hall effect measurements in this regard. Such measurements have been made on $RbAg_4I_5$ [3.29], but the interpretation of the results does not seem to shed any new light on the matter (see Sect. 3.1.4).

$Py_5Ag_{18}I_{23}$

The other solid electrolyte in the PyI-AgI system is of particular importance because it is a "two-dimensional" solid electrolyte, as will soon be demonstrated. $Py_5Ag_{18}I_{23}$ belongs to space group $P\bar{6}2m$ (D_{3h}^3) with lattice constants $a = 13.62 \pm 0.02$, $c = 12.58 \pm 0.02$ Å, and the unit cell contains one formula unit. The structure, determined by *Geller* et al. [3.30, 31], is depicted in Fig. 3.5.

In this structure each unit cell contains 23 I^- ions, 22 of which form 55 tetrahedra which share faces in such manner as again to provide pathways for the motion of the Ag^+ ions. However, these pathways are confined to slabs separated by layers of three Py^+ ions and one I^- ion per unit cell. There are no paths penetrating these layers (see Fig. 3.5). From a knowledge of this structure, it was stated [3.30, 31] that the electrolytic conductivity in the c-axis direction is essentially zero; this has now been confirmed [3.24].

In the $Py_5Ag_{18}I_{23}$ structure, the Ag^+ ions partially occupy seven sets of crystallographically nonequivalent sites, one 3-fold, one 4-fold, two 6-fold and three 12-fold sets. Again, the distribution of Ag^+ ions over these sets is markedly nonuniform. Indeed one of the 12-fold sets is empty within experimental error.

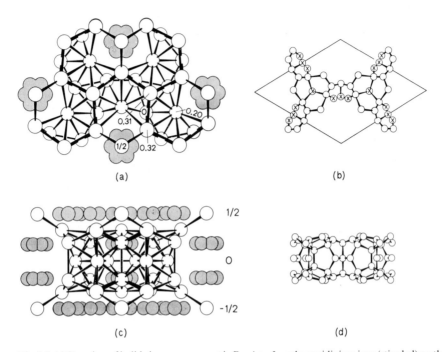

Fig. 3.5. (a) Top view of iodide ion arrangement in $Py_5Ag_{18}I_{23}$; the pyridinium ions (stippled) on the hexagonal axes are also shown. (b) Top view of Ag^+ ion paths in $Py_5Ag_{18}I_{23}$; the equilibrium Ag^+ ion sites (located at or near iodide tetrahedra centers) are shown; × indicates connection between upper and lower halves of the Ag^+ path network within the conducting layer; see also (d). (c) Side view of the iodide arrangement in $Py_5Ag_{18}I_{23}$; the pyridinium ions (stippled) in the $\pm 1/2c$ levels of the unit cell are shown; these together with the I^- ions at $\pm(0, 0, 1/2)$ block movement of Ag^+ ions in the c-axis direction. (d) Side view of the Ag^+ ion paths in $Py_5Ag_{18}I_{23}$. (From [3.31]; reprinted by permission of the publisher, The Electrochemical Society, Inc.)

Strictly on a "numbers basis", it is possible for the Ag^+ ions to become ordered in this structure: the Ag^+ ions could completely fill one of the 12-fold and one of the 6-fold sets. *Hibma* [3.24] has made measurements of the conductivity down to a temperature of 160 K and specific heat measurements down to 100 K, without finding any indicating of a phase change. There is nevertheless a possibility that ordering of the Ag^+ ions may take place at a yet lower temperature. It is to be noted that two of the five Py^+ ions per unit cell are disordered in the same manner as all the Py^+ ions in the hexagonal phases of $PyAg_5I_6$.

Silver Iodide Tetratungstate, $Ag_{26}I_{18}W_4O_{16}$

The structure of $Ag_{26}I_{18}W_4O_{16}$ [3.32] is one in which substitution is made for iodide rather than for Ag^+ ions. This solid electrolyte was first reported by *Takahashi* et al. [3.33] to have the formula $Ag_6I_4WO_4$.

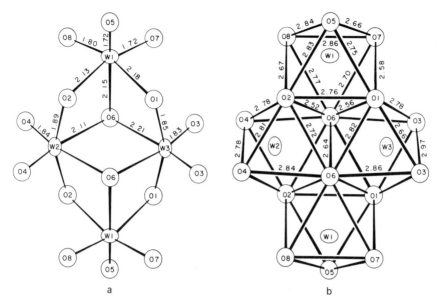

a   b

Fig. 3.6a and b. The tetratungstate ion, $(W_4O_{16})^{8-}$, in $Ag_{26}I_{18}W_4O_{16}$. (From [3.32].) Note: this ion is actually the enantiomorph of that found in $Ag_8W_4O_{16}$

As a first step it was necessary to determine what kind of tungstate entity to expect in the solid electrolyte. Therefore, the structure of the high temperature phase of $Ag_8W_4O_{16}$ was determined [3.34]. The tungstate complex in this phase is a tetratungstate ion, $(W_4O_{16})^{8-}$, a sketch of which is shown in Fig. 3.6. This ion has only twofold symmetry and is apparently the basis for the symmetry of crystals of $Ag_{26}I_{18}W_4O_{16}$. To some extent the symmetries of the cation-substituted crystals are also based on the symmetries of the substituent cations. In the $Ag_{26}I_{18}W_4O_{16}$ structure, the influence of the $(W_4O_{16})^{8-}$ ion is more pervasive than, for example, that of the pyridinium ion in $PyAg_5I_6$ or of the $[(CH_3)_4N]^+$ ion in $[(CH_3)_4N]_2Ag_{13}I_{15}$. The symmetry of the free $[(CH_3)_4N]^+$ ion is $\overline{4}3m$; the rhombohedral structure of $[(CH_3)_4N]_2Ag_{13}I_{15}$ makes use only of one of the threefold axes, and in this structure the symmetry of the $[(CH_3)_4N]^+$ ion is only 32.

Crystals of $Ag_{26}I_{18}W_4O_{16}$ belong to space group C2 (C_2^3) and contain $2\,Ag_{26}I_{18}W_4O_{16}$ in the monoclinic unit cell with $a = 16.76 \pm 0.03$, $b = 15.52 \pm 0.03$, $c = 11.81 \pm 0.02$ Å, $\beta = 103.9 \pm 0.1°$. The crystal structure of $Ag_{26}I_{18}W_4O_{16}$ [3.32] is rather difficult to describe briefly, but some features will be given here.

There is a total of 90 iodide polyhedra in $Ag_{26}I_{18}W_4O_{16}$, 88 tetrahedra and two octahedra per unit cell, which share faces and provide a three-dimensional network of passageways through which the Ag^+ ions move. Surrounding each tetratungstate ion there are 60 polyhedra, therefore 120 per unit cell, which are formed from both I^- ions and O atoms. Eight of these are octahedra which are

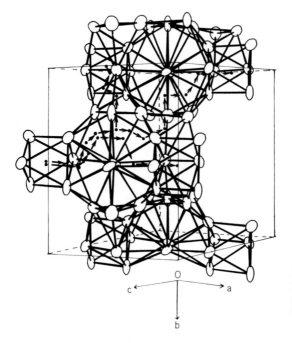

Fig. 3.7. The (pure) iodide arrangement in $Ag_{26}I_{18}W_4O_{16}$. Some pathways for Ag^+ ions are shown. (From [3.32])

fully occupied by Ag^+ ions. The remaining 112 I–O are either tetrahedra or five-cornered polyhedra. Of these, 12 are vacant because they are too close to tungsten atoms; 16 are vacant because they are too close to the filled octahedra, and 28 which have rather high occupancies are not in conduction passageways. These 28 sites contain Ag^+ ions which may be in *static* disorder. The remaining 56 I–O polyhedra share faces with the iodide polyhedra and *are* in conduction passageways.

The total number of mobile Ag^+ ions (that is, those in the conduction passageways) is 23.2 or 44.5 % of the Ag^+ ions in the unit cell. The remainder are in filled I–O octahedra, or in I–O polyhedra that are not in conduction passageways. The occupancies of the I–O polyhedra in the conduction passageways are in general higher than those of the iodide polyhedra, indicating that the mobilities of Ag^+ ions through the former are lower than those through the latter. $Ag_{26}I_{18}W_4O_{16}$ is a three-dimensional solid electrolyte.

The original paper on this structure contains a detailed description (including seven figures) of the structure and the conduction passageways. One of the figures showing only the iodide arrangement and some routes Ag^+ ions can take is reproduced in Fig. 3.7. The interested reader is referred to the original paper for further information.

α-Silver Iodide, α-AgI

It might be thought that the description of α-AgI should have come first. However, there is good reason for leaving it to this point. *Tubandt* and *Lorenz*

[3.35] were the first to observe the important transition at 419 K to this highly conducting phase. The high conductivity of 1.31 $[\Omega\,cm]^{-1}$ occurs right at the transition.

About twenty years after this discovery, *Strock* [3.36, 37], working with x-ray powder diffraction photographs, found that α-AgI belongs to space group Im3m (O_h^9), the cubic unit cell, $a = 5.044$ Å, containing two AgI, with iodide ions at the corners and body center.

Strock concluded that the Ag$^+$ ions were distributed uniformly over 42 space group sites: 6b, 12d, and 24h. The 6b sites *appear* to be octahedrally coordinated, but they can more correctly be considered to have twofold coordination to two I$^-$ ions at 2.52 Å. The other four distances are 3.57 Å, which are quite long. Thus the 6b sites are high energy sites, and it is unlikely that Ag$^+$ ions would reside there long if they move through such sites at all. The 12d sites are at the centers of tetrahedra which share faces and provide a network of pathways through which the Ag$^+$ ions can move. The 24h sites are at the centers of the shared faces, and the Ag$^+$ ions are expected to move rapidly through them.

The reason for discussing α-AgI at this point can now be made clear. As a result of the structural model explaining the electrolytic conductivity in RbAg$_4$I$_5$ [3.1], *Wiedersich* and *Geller* [3.17] suggested that in α-AgI, the conduction passageways for Ag$^+$ ions are formed by the face-sharing tetrahedra. This is supported by the results of *Bührer* and *Hälg* [3.38] and of *Cava* and *Wuensch* [3.39], namely, that the Ag$^+$ ions are confined to the iodide tetrahedra.

Rahlfs [3.40] reported that the structures of α-Ag$_2$S and α-Ag$_2$Se are of a type similar to the *Strock* α-AgI structure. However, a preference of the Ag$^+$ ions for the tetrahedral sites was indicated in α-Ag$_2$Se. *Hoshino* [3.41] has reported that α-CuBr has the *Strock* α-AgI structure. It is to be expected that a single-crystal investigation will show results that are analogous to those for α-AgI.

Silver Sulfide Halogenides, Ag$_3$SI and Ag$_3$SBr

These materials were reported by *Reuter* and *Hardel* [3.42–44] (see also [3.45–47]). Ag$_3$SI undergoes a crystallographic transition at 235 °C. In the β-phase and in Ag$_3$SBr, the S^{2-} and I$^-$ ions are completely ordered, one at the body-center the other at the corners of a cubic unit cell; in the α-phase the S^{2-} and I$^-$ ions are uniformly distributed over the body-centers and corners. In both cases the Ag$^+$ ions are distributed over 12-fold tetrahedral sites corresponding to the distribution in α-AgI. The α-phase has an order of magnitude higher conductivity than the β-phase at the transition [3.42, 44]. The volume of the unit cell increases by 5.6%. The number of carriers per unit cell remains the same (although because of the increased volume the carrier *concentration* decreases by 5.6%). To *increase* the conductivity by an order of magnitude, the mobility of the Ag$^+$ ions must increase markedly. It has been pointed out [3.17] that the "squeezing" of the Ag$^+$ ions in passing from one tetrahedral site to the next most likely determines the activation energy of motion, h_m. The squeezing is defined as the difference between the radius of a sphere passing through the vertices of the tetrahedron and that of the circle passing through the vertices of a face. The values of h_m for β-Ag_3SI and Ag_3SBr are 0.21 and 0.28eV [3.44], respectively, while that for α-Ag_3SI is only 0.08 eV [3.44]. Thus it appears that the squeezing parameter must be smaller for α-Ag_3SI than for β-Ag_3SI or Ag_3SBr. (It should be pointed out that the squeezing parameters are not so easily defined for the less symmetric structures.) Another possibility is that the disorder of the anions might favor increased availability of octahedral sites for the Ag^+ ions in the α-phase.

α-Silver Mercuric Iodide, α-Ag_2HgI_4

α-AgI, α- and β-Ag_3SI, α-Ag_2S, and α-Ag_2Se have structures in which the anions are in a "body-centered" arrangement [3.17]. Another anion arrangement found in some solid electrolytes is face-centered [3.17]. This is the case of α-Ag_2HgI_4, α-Cu_2HgI_4, [3.48, 49] α-Ag_2Te, α-Cu_2Se, α-$Cu_{1.8}S$ [3.40], α- and γ-CuI [3.50]. The space group is $F\bar{4}3m$. The cubic unit cell of α-Ag_2HgI_4 has $a = 6.38$ Å at 50 °C and contains one Ag_2HgI_4. *Ketelaar* [3.48, 49] suggested that the three cations are uniformly distributed over the 4d sites, those that are occupied by Zn^{2+} ions in zincblende or by the Ag^+ ions in γ-AgI.

In this structure, iodide tetrahedra share only corners with other tetrahedra, iodide octahedra share corners and edges with other octahedra, but octahedra share faces with tetrahedra. The pathways for the cations must therefore consist of alternating tetrahedral and octahedral sites [3.17]. The crystal structure of α-Ag_2HgI_4 has recently been reinvestigated and refined from single crystal x-ray diffraction data by *Kasper* and *Browall* [3.51]. It was found that the structure proposed by *Ketelaar* is essentially correct with regard to the distribution of the cations. *Kasper* and *Browall*, however, included the effect of anharmonicity of the cation motions in their calculations, which gave better agreement with the observed data than does the *Ketelaar* model involving only harmonic cation motions.

The structure of α-Ag_2HgI_4 is closely related to that of γ-AgI, which has the ordered zincblende structure, and is not a true solid electrolyte. Thus the presence of a divalent ion and/or a vacant site destabilizes the zincblende-type structure and makes it a solid electrolyte, though not a particularly good one. An interesting feature is that only one symmetry related set of tetrahedral sites contains cations. The other set of tetrahedral sites and the octahedral sites are found by the x-ray investigation to be empty, which means that either the cations move through them rapidly, or (as is possible in this case) not at all.

Browall and *Kasper* [3.52] have measured the conductivity of single crystals of α-Ag_2HgI_4 and obtained a value of 0.33 eV for the enthalpy of activation of motion, a value slightly smaller than that obtained by *Ketelaar* [3.49] from measurements on polycrystalline material. This is still larger than the values, 0.1–0.2 eV, found for good solid electrolytes. *Browall* and *Kasper* suggest that this high activation enthalpy is associated with the creation of interstitial ions in the octahedral sites. They point out that this together with the occupancy of only one set of tetrahedral sites (and no appreciable occupancy of any others) and the large anharmonic vibrations in the direction of the octahedral sites supports the *Wiedersich-Geller* [3.17] model for conductivity of α-Ag_2HgI_4 (see above).

Ketelaar [3.53] has shown that the transport number of the Hg^{2+} ions is only 0.06. Thus most of the current is carried by the Ag^+ ions. This may be the reason for the rather high *average* activation enthalpy of motion obtained for α-Ag_2HgI_4.

As a result of a diffuse x-ray scattering investigation made on α-Ag_2HgI_4, *Hibma* et al. [3.54] suggest that the discrepancies between observed and calculated intensities found for some Bragg reflections by *Kasper* and *Browall* [3.51] are caused by static disorder rather than by anharmonicity of the ionic motions. They suggest that a realistic model for this phase is the "dispersion of small 'ordered' domains in a disordered matrix". It seems that the exact nature of α-Ag_2HgI_4 is not yet resolved.

3.1.2 Miscellaneous Compounds Based on AgI

Numerous solid electrolytes based on AgI have been found on which crystal structure work has not yet been done. These include compounds with 1) tetraalkylammonium iodides [3.3, 19]; with 2) saturated and 3) unsaturated azacyclic substituted ammonium iodides [3.3, 55]; with 4) saturated carbocyclic, 5) aryl, 6) benzyl and 7) allyl ammonium iodides [3.55]; with 8) azonia ions containing one or more heteroatoms [3.56]; and with 9) sulfonium ions [3.57]. The maximum conductivities at 22 °C are 0.06, 0.06, 0.04, 0.02, 0.008, 0.01, 0.06, 0.05, and 0.06 $[\Omega\,cm]^{-1}$, respectively. (Please note that there are three subsets of compounds in [3.56].) It should be emphasized that it is possible that the conductivity reported is low if the material is not single phase. For example, the value reported [3.3, 55] for $PyI \cdot 8\,AgI$ is 0.04 $[\Omega\,cm]^{-1}$; the correct formula is $PyAg_5I_6$, which has a conductivity of 0.077 $[\Omega\,cm]^{-1}$ at room temperature [3.22].

Berardelli et al. [3.58] have reported AgI-based solid electrolytes involving polymethonium diiodides (12:1 ratio) with conductivities in the range 0.01 to 0.05 $[\Omega\,cm]^{-1}$. They have also reported that the compound formed from octamethyldiethyltriammonium triiodide and silver iodide in 1:22 ratio has a room temperature specific conductivity of 0.058 $[\Omega\,cm]^{-1}$.

In addition to the compound involving silver tungstate and silver iodide, *Takahashi* and coworkers have reported the solid electrolytes $Ag_7I_4PO_4$ and $Ag_{19}I_{15}P_2O_7$ with (25 °C) specific conductivities, 0.02 and 0.09 $[\Omega\,cm]^{-1}$, respectively [3.59], a solid electrolyte which is a solid solution of 22 mol-% Ag_2SO_4 in α-AgI, which has a room temperature specific conductivity of 0.05 $[\Omega\,cm]^{-1}$ [3.60], and the solid electrolytes $Ag_2Hg_{0.25}S_{0.5}I_{1.5}$, $Ag_{1.85}Hg_{0.40}Te_{0.65}I_{1.35}$, $Ag_{1.80}Hg_{0.45}Se_{0.70}I_{1.30}$, and $Ag_{2.0}Hg_{0.5}Se_{1.0}I_{1.0}$ with conductivities, at 25 °C, of 0.15, 0.09, 0.10, and 0.04 $[\Omega\,cm]^{-1}$ respectively [3.61]. The last groups are said to have crystal structures related to that of α-AgI.

Takahashi and *Yamamoto* have also reported [3.62] a body-centered cubic phase $Ag_2Hg_{0.4}S_{0.6}I_{1.6}$ with lattice constant $a = 4.96$ Å that has a conductivity of 0.04 $[\Omega\,cm]^{-1}$ at 25 °C. Earlier, *Suchow* and *Pond* [3.63] studied the photosensitive and phototropic properties of products of solid state reaction in the Ag_2S-HgS system but did not measure the conductivities of their materials.

Coetzer and *Thackeray* [3.64] prepared double salts of AgI and alkyldiamine diiodides; results of conductivity measurements have been reported on two of these, one having the rather high value of 0.11 $[\Omega\,cm]^{-1}$ at 22 °C.

Christie et al. [3.65] have reported conductivity results on the tropyllium iodide-silver iodide system. A maximum of $5.8 \cdot 10^{-3}$ $[\Omega\,cm]^{-1}$ occurs at 80–85 mol-% AgI; the exact formula of the compound could not be determined from the data.

3.1.3 Relation of Conductivity to Structure

We have seen in the above sections that many solid electrolytes are formed between AgI and various iodides and silver compounds. In all cases (for which structure determinations have been made) the solid electrolytes have certain structural features in common which we shall summarize below. First we discuss briefly the role of the immobile substituent ions, which are usually quite large relative to the Ag^+ ions.

The structures demonstrate that the role of the substituent ions is to stabilize structures in which mainly iodide tetrahedra share faces rather than corners as in γ-AgI. This generally results in bringing adjacent sites closer together than in a corner-sharing structure. This appears, at first hand, to be a condensation and in certain respects it is.

A unit cell of γ-AgI contains eight tetrahedral and four octahedral sites in a volume of 274 Å^3. A unit cell of α-AgI contains twelve tetrahedral sites in 128 Å^3. Even though the volume of the octahedron is four times that of the tetrahedron, it is nevertheless true that the *site concentration* in the latter is substantially higher than that of the former. In $RbAg_4I_5$, the volume attributed to the 56 tetrahedra is 669 Å^3 [3.31] or, on the average, 143 Å^3 for 12 tetrahedra. However, more than half the crystal space is attributable to the four octahedra surrounding the Rb^+ ions and unoccupiable interstices. In general, the substitutent immobile cations and the parts of the iodide ions that are considered to belong to them occupy approximately half or more of the crystal space (Table 3.2).

The average coordination of the ions does not change significantly in the solid electrolytes relative to the separate salts, and therefore the volumes occupied are not much different. For example, the volume per RbI in RbI is 98.9 Å^3 and the volume per AgI from γ-AgI is 68.2 Å^3. The total volume of 4 RbI and 16 AgI is 1486.8 Å^3 as opposed to the volume 1420 Å^3 of a unit cell of $RbAg_4I_5$. If we take the volume per AgI from α-AgI at 146 °C, i.e., 64.2 Å^3, we obtain for the total volume 1422 Å^3, within experimental error, equal to that of the unit cell of $RbAg_4I_5$. It seems strange, but it is nevertheless true, that we

Table 3.2. Some specialized data on AgI-based solid electrolytes

Formula	Mobile Ag^+ ions/unit cell	Ag^+ ion sites/unit cell	$\dfrac{\text{Vol. of channels}}{\text{Vol. of unit cell}}$	n_{Ag} 10^{22} cm^{-3}	n_s 10^{22} cm^{-3}	$\sigma\,[\Omega\,\text{cm}]^{-1}$ 296K	419K
α-AgI	2	12	1.00	1.57	9.42	—	1.3
$RbAg_4I_5$	16	56	0.471	1.13	3.94	0.27	0.66
$PyAg_5I_6$	10	34	0.525	1.07	3.65	0.077	1.1
$[(CH_3)_4N]_2Ag_{13}I_{15}$	13	41	0.39	1.04	3.28	0.04	0.26
$Ag_{26}I_{18}(W_4O_{16})$	23.2	$90^a + 56^b$	$0.344^a + 0.221^b$	0.78	4.96	0.058	0.31
$Py_5Ag_{18}I_{23}$	18	55	0.316	0.89	2.72	0.008	0.05

[a]   Pure iodide polyhedra.
[b]   Mixed I–O polyhedra.
Note: $n_{Ag} \equiv Ag^+$ ion concentration; n_s = site concentration.

cannot get such agreement from calculations of the volumes of the polyhedra. For α-AgI, the volume of an iodide tetrahedron is 10.7 Å^3, while the volume of an RbI octahedron in RbI is 131.8 Å^3, both much smaller than the respective polyhedra in $RbAg_4I_5$. Also in $RbAg_4I_5$, there are interstices which cannot be occupied, but even if account is taken of their volumes, the total is still low; in $RbAg_4I_5$ the volumes of the polyhedra are larger than in the separate compounds.

Another example is that of $PyAg_5I_6$. The volume per PyI [3.26] is 165.2 Å^3. The separate formula unit volumes give a total of 1012 Å^3 or 972 (from α-AgI) to be compared with 931 Å^3 for the unit cell of $PyAg_5I_6$. However, in this case, the average coordination of both the Py^+ and Ag^+ ions is increased over the respective coordination in the separate compounds. The face-sharing of the polyhedra also tends to reduce the volume relative to corner sharing; at 146 °C the density of α-AgI is 6.07 g/cc while that of γ-AgI at 25 °C is 5.68 g/cc. In the case of $Ag_{26}I_{18}W_4O_{16}$, the totals from separate volumes for 2 $Ag_8W_4O_{16}$ and 36 AgI are 3231 and 3087 (AgI from α-AgI) Å^3 whereas the volume of the monoclinic cell is 2982 Å^3 [3.32].

It appears that a reaction of AgI with another compound will be favored if it results in lower average molar volumes for the components. However, such reactions are not predictable. The work in obtaining the solid electrolytes has actually been based on a crystal chemical approach, which is after all the one usually used successfully in obtaining new materials. As *Owens* has pointed out [3.19, 55, 65] the ionic size of the substituent ion plays an important role. It seems also, as indicated in Section 3.1.1, that the symmetry of the substituent ion may play an important role in determining the nature of the iodide framework. In the $Ag_{26}I_{18}W_4O_{16}$ case, the oxygens are actually involved in the formation of the conduction pathways and "trapping" sites as they must; inasmuch as there are only 90 iodide face-sharing polyhedra and 52 Ag^+ ions in the unit cell, there are not enough of the former to accommodate the latter even in an ordered structure.

It has already been emphasized in Section 3.1.1 that the AgI-based solid electrolytes all have networks of passageways formed from the face-sharing of the anion polyhedra. This has now been discussed in many papers [3.1, 15, 17, 21–23, 30–32, 66, 67]. This feature occurs in almost all solid electrolytes of any kind and will be shown to be the case in subsequent sections of this chapter and is to some extent discussed in other chapters.

Certain points are worth emphasizing here. 1) In structures in which current carrier sites are not crystallographically equivalent, the distribution of the current carriers over the different sites is markedly *nonuniform*. This is equivalent to the statement that crystallographically nonequivalent sites have different site energies. 2) As long as the crystal structure does not change, it is highly probable that the distribution of current carriers over crystallographically nonequivalent sites will remain nonuniform up to the melting point, but the distributions do change with temperature [3.17, 22]. 3) The conductivity appears to be associated with the nature of the passageways; the simpler the latter are, the higher will be the conductivity. 4) Three-dimensional networks give higher average conductivities than two-dimensional networks. This is particularly illustrated by a comparison of $Py_5Ag_{18}I_{23}$ with $PyAg_5I_6$ and other three-dimensional cases. 5) Higher site concentrations give higher conductivities. 6) Larger volumes of crystal space occupied by the conduction passageways lead to higher conductivities. This is related to both points 4 and 5. 7) The ratio of available sites to current carriers is important. A factor of 2 will probably not give a solid electrolyte, because it is probable that at least half the sites will be too close to the other half. The highest factors (Table 3.2) in the AgI-based solid electrolytes are for α-AgI itself and *apparently* for $Ag_{26}I_{18}W_4O_{16}$ (Table 3.2). The lowest (Table 3.2) is for $Py_5Ag_{18}I_{23}$, which has a low conductivity. 8) Interactions between cations on nearest neighbor tetrahedral sites must be strong and repulsive [3.17, 18, 24, 68], because the intersite distances are so short as to imply that such sites cannot be occupied simultaneously. Interactions on nearest neighbor octahedral-tetrahedral sites as in $PyAg_5I_6$ must be much weaker, else $PyAg_5I_6$ could not be a solid electrolyte (see Sect. 3.1.1). 9) The "squeezing" of the Ag^+ ions on passing through a shared triangular face from one polyhedron to another probably determines the activation enthalpy of motion [3.17]. This is essentially the cation-anion interaction to which *Pardee* and *Mahan* [3.18] refer (see also Sect. 2.2). 10) According to *Armstrong* et al. [3.69], the stability of the Ag^+ (and Cu^+) ions in both 4- and 3-coordination and their monovalency are responsible for their being the mobile ion in most of the good solid electrolytes.

Some of the above listed points are illustrated in Table 3.2. Both room temperature and 419 K average conductivites are given, the latter to facilitate comparison with α-AgI. Several of the conductivities are extrapolated values, but the conductivities of the six solid electrolytes listed have been measured to at least 375 K. The conductivities are averages over the polycrystalline materials. (For $Py_5Ag_{18}I_{23}$ at 295 K, $\sigma_1 = \sigma_2 = 0.012$ $[\Omega\,cm]^{-1}$, $\sigma_3 \approx 0$ [3.31]; $\sigma_1 = \sigma_2 = 0.017$ $[\Omega\,cm]^{-1}$, $\sigma_3 \approx 10^{-5}$–10^{-6} $[\Omega\,cm]^{-1}$ [3.25]; for $PyAg_5I_6$ at 295 K, $\sigma_1 = \sigma_2 = 0.04$ $[\Omega\,cm]^{-1}$, $\sigma_3 = 0.17$ $[\Omega\,cm]^{-1}$ [3.25].) Because α-AgI and

$RbAg_4I_5$ are cubic, the conductivities are, of course, isotropic, so that the values for the polycrystalline materials should be equal to those of the single crystals (neglecting effects of grain boundaries).

While the points listed above seem to hold for the compounds listed in Table 3.2, the relations are by no means simple. For $PyAg_5I_6$ and $RbAg_4I_5$ the carrier concentrations are nearly the same, but the site concentration of the former is larger than that of the latter; yet at 419 K the conductivity of the former is almost twice that of the latter. In fact, the conductivity of $PyAg_5I_6$ is nearly equal to that of α-AgI at 419 K.

It has been predicted [3.22, 67] and has now been confirmed [3.24] that the conductivity of $PyAg_5I_6$ is highly anisotropic; the straight channels along the c-axis, formed by the face-sharing octahedra, are much more simply traversed by the Ag^+ ions than any other pathways. $PyAg_5I_6$ has only about two-thirds the carrier concentration of α-AgI and little more than half its volume given to passageways for the Ag^+ ions as opposed to 100% for α-AgI, but the latter does not have the simple channels that the former has along its c-axis.

The carrier and site concentrations of $[(CH_3)_4N]_2Ag_{13}I_{15}$ are 16 and 20% higher than those of $Py_5Ag_{18}I_{23}$, but the conductivity of the former is five times that of the latter. The lowest conductivity of the six is that of $Py_5Ag_{18}I_{23}$ which has the lowest site concentration (or passageway volume) of all. Doubtless the low conductivity of $Py_5Ag_{18}I_{23}$ is directly attributable to its being a two-dimensional conductor.

According to our assumptions [3.32], the *mobile* carrier concentration in $Ag_{26}I_{18}W_4O_{16}$ is $0.79 \cdot 10^{22}$ cm^{-3}, the lowest of the six listed in Table 3.2. Its site concentration is high relative to the other solid electrolytes. Its conductivity, however, is comparable with that of $[(CH_3)_4N]_2Ag_{13}I_{15}$, even though the site concentration of the latter appears to be much smaller. This must be related to the rather less favorable arrangement of the passageways in $Ag_{26}I_{18}W_4O_{16}$ and the somewhat higher carrier concentration in $[(CH_3)_4N]_2Ag_{13}I_{15}$. It should also be kept in mind that it is probable that the enthalpies of activation of motion out of the sites surrounded by mixed I–O polyhedra are higher than for the pure iodide polyhedral sites. This is deduced from the generally substantially higher occupancies of the I–O polyhedra relative to the pure iodide polyhedra [3.32].

(A paper relevant to the conductivity in these as well as in other solid electrolytes is by *Heyne* [3.70].)

3.1.4 Other Results on $RbAg_4I_5$ and KAg_4I_5

Many measurements have been made on $RbAg_4I_5$ to various purposes. *Kaneda* and *Mizuki* [3.29] have measured the Hall effect. In the dark, the Hall mobility at room temperature is 0.05 cm^2 V^{-1} s^{-1}, and the Hall voltage is opposite in sign to that obtained when the crystal is illuminated with light from a high-pressure mercury lamp. This observation together with the relative magnitudes and temperature dependence of the Hall mobility imply that the Hall signals in the dark are caused by the Ag^+ ion carriers. Under illumination, the primary contributors to the Hall voltage are the photoelectrons.

The authors use the *Emin-Holstein* [3.71] small polaron model to calculate the Ag^+ ion mobility for comparison with the measured value. They also use an average jump distance of $[11.4/(56)^{1/3}] \cdot 10^{-8}$ cm $= 2.98 \cdot 10^{-8}$ cm, for which there is really no justification; the structure analysis [3.1] shows that the jump distances are 1.68, 1.71, 1.76, and $1.91 \cdot 10^{-8}$ cm with a weighted average of approximately $1.75 \cdot 10^{-8}$ cm. With the latter average jump distance and the other assumptions made by the authors, the calculated Ag^+ ion mobility [from their Eq. (1)] is 0.008 cm^2 V^{-1} s^{-1}, only one-sixth of their observed value. The authors also calculated the conductivity according to the *Emin-Holstein* theory and obtained a result which is an order of magnitude higher than the observed value. From this they conclude that only about 10% of the Ag^+ ions are effective carriers. It seems probable that the application of the *Emin-Holstein* model to the $RbAg_4I_5$ case is an oversimplification.

Bentle [3.72] has measured the silver ion diffusion in $RbAg_4I_5$. The magnitudes of the diffusion coefficients were found to be large relative to those reported for any other material. The activation energy for diffusion was found to be 0.084 eV.

Nagao and *Kaneda* [3.73] have reported the results of ultrasonic attenuation measurements by single crystals of $RbAg_4I_5$. They found that the activation energies of motion of the Ag^+ ions lie between 0.06 and 0.12 eV, close to the value of 0.10 eV found [3.5, 7] from conductivity measurements and 0.084 eV found [3.72] from diffusion measurements.

Thermoelectric power measurements have been made on $RbAg_4I_5$ by *Johnson* et al. [3.74] and by *Danilov* et al. [3.75], and on KAg_4I_5 by *Chandra* et al. [3.76] and by *Johnson* et al. [3.74]. *Johnson* et al. [3.74] found the heat of transport of the Ag^+ ions in solid $RbAg_4I_5$ to be 2.1 kcal mole^{-1} or 0.09 eV, in good agreement with the activation enthalpy obtained from conductivity and tracer-diffusion measurements. *Danilov* et al. [3.75] made their measurements in the temperature range $-100°$ to 400 °C and found an essentially constant value $-590 \pm 15\,\mu$V K^{-1} for the thermoelectric power. The values found by *Johnson* et al. in the temperature region 80° to 180 °C are in the range $-530 \pm 30\,\mu$V K^{-1}.

By means of wavelength-modulated absorption and reflectance measurements, *Bauer* and *Huberman* [3.77] have studied the effect of the cation disorder on the electronic states of α-$RbAg_4I_5$. Compared with β-AgI which is not a solid electrolyte, they found that the cation disorder causes a broadening of the valence-band edge. The results include a proposed room-temperature band structure for $RbAg_4I_5$ with a direct gap of 3.3 eV.

Burns et al. [3.78] have found that γ-$RbAg_4I_5$ is piezoelectric. This is in accord with its belonging to point group 32 [3.15]. (β-$RbAg_4I_5$ should also be piezoelectric because it belongs to the same point group [3.15], but α-$RbAg_4I_5$ belongs to point group 432 for which all piezoelectric moduli vanish.)

Raman spectroscopic studies have been made on $RbAg_4I_5$ by *Burns* et al. [3.78] and by *Delaney* and *Ushioda* [3.79]. The results of the measurements by the two groups are in agreement; the former authors, however, give a more definitive discussion of the results. Of particular interest are the observations that no changes in the Raman spectra could be detected at the 209 K transition and that at the 122 K transition the only change is the abrupt disappearance of a small phonon mode at 22.9 cm^{-1} in the β-phase. This is contrasted with the case of AgI [3.80] for which there are considerable discontinuous changes in the Raman spectrum at the 147 °C transition. The authors conclude that the structural changes of RbAg$_4$I$_5$ are minor relative to those of AgI; this is in accord with the results of the crystal structure investigation of the low temperature phases (see Sect. 3.1.1).

3.2 Solid Electrolytes Involving the Cu$^+$ Ion

The replacement of Ag$^+$ by Cu$^+$ in solid electrolytes is of both scientific and practical interest, the latter because copper is so much less expensive than silver. Needless to say, from a practical point of view, one may expect a problem of stability of the Cu$^+$ ion in the presence of oxygen that does not exist in the case of Ag$^+$. There is also a significant difference in the sizes of the two ions, and this could be the reason that there are very few Cu$^+$ ion solid electrolytes that are isostructural with Ag$^+$ ion solid electrolytes.

Some binary cuprous halogenides and KCu$_4$I$_5$ have been discussed in Section 3.1.1. *Takahashi* et al. [3.81] have reported five compounds involving an N-alkyl (or hydro) hexamethylenetetramine halide with a cuprous halide: $(C_6H_{12}N_4CH_3)_3Cu_{17}I_{20}$, $(C_6H_{12}N_4H)Cu_7Br_8$, $(C_6H_{12}N_4CH_3)Cu_7Br_8$, $(C_6H_{12}N_4C_2H_5)Cu_7Br_8$, and $(C_6H_{12}N_4H)Cu_7Cl_8$ with electrical conductivities at 20 °C of 0.001, 0.006, 0.02, 0.03, and 0.04 [Ω cm]$^{-1}$ respectively. *Sammells* et al. [3.82] have reported eight Cu$^+$ solid electrolytes with a conductivity range at 25 °C of 0.002–0.02 [Ω cm]$^{-1}$. These are either bromides or chlorides involving the ions of quinuclidine, 3-quinuclidone, pyridinium, N,N'-dihydro-triethylenediamine and N,N' dimethyltriethylenediamine.

3.3 Solid Electrolytes with Fluorite Structure

Several halogenide compounds with the cubic fluorite (CaF$_2$) structure (Fig. 6.1) are solid electrolytes at relatively high temperatures. One of these is CaF$_2$ itself [3.83, 84]; others are SrF$_2$, BaF$_2$ [3.84, 85], SrCl$_2$ [3.86], BaCl$_2$, and SrBr$_2$ [3.84, 86]. The last two do not have the fluorite structure in their normal state; they go through first order transitions at which their conductivities rise steeply. Their high temperature structures are predicted [3.85] to be fluorite type. The high temperature form of PbF$_2$ is also a fluorite-type solid electrolyte [3.87, 88]. In all these cases, the anion is the current carrier.

In the fluorite-type structure (Fig. 6.1), each cation is surrounded by a cube of anions and each anion is surrounded by a tetrahedron of cations. The cations are located at the corners and face-centers of the cubic unit cell; therefore the anion cubes surrounding the edge-centers and body center are empty. It is presumably the availability of these sites which makes the fluorite-type crystals solid electrolytes. Each anion is at the corner of eight cubes, half of which are empty. Thus when an anion is thermally excited into a formerly empty ("interstitial") site, eight of the anion cubes are left with only seven anions at their corners. It would seem that when this occurs there would be a tendency for the anions in such "defect" cubes to be drawn closer to the cation and that the anions surrounding the anion in its defect cube to be repelled from it, the two tendencies being consistent with each other. There are also forces that tend to counteract this: neighboring cubes that are intact, for example.

A second anion from the first cube might also be excited into a neighboring interstitial site in which case it would leave the first cube with two anions missing and seven others with only one anion missing. This would increase the tendency toward distortion of the surroundings of the cation and possibly cause displacement of the cation itself. The situation could be rather complex at high temperatures and calls for a single crystal diffraction investigation. (But see also [3.89] and Chap. 6.)

Derrington and *O'Keeffe* [3.86] suggest that cooperative modes of motion (rather than isolated jumps) must be involved in transport in these materials as had been suggested by *Funke* and coworkers [3.90, 91] for the cation conducting α-RbAg$_4$I$_5$ and α-AgI. In subsequent papers [3.92, 93] *Funke* and coworkers present as their model the superposition of local motions on the jumps, with the mean residence time (at a site) comparable to the mean jump-duration time.

The gradual increase of conductivity with increasing temperature of the fluorite-type solid electrolytes appears to be analogous to the PyAg$_5$I$_6$ case (Sect. 3.1.1). Consider CaF$_2$, on which substantial work has been done. The gradual increase in its conductivity is accompanied by an increase in the concentration of F$^-$ ions in the interstitial sites and an equal increase in the F$^-$ ion or tetrahedral site vacancy concentration, causing a significant change in the crystal structure. (See the discussion on PyAg$_5$I$_6$ in Sect. 3.1.1.)

Naylor's [3.94] heat content data on MgF$_2$ and on CaF$_2$ were replotted as heat capacity data by *Dworkin* and *Bredig* [3.95] in a paper on the heat content of K$_2$S. The heat capacity data indicate a λ-type transition with a broad tail on the low temperature side, as one might expect for a gradual disordering process. The vertical part of the λ is at 1424 K. Transitions are also observed for SrCl$_2$ and for K$_2$S (see Fig. 3.8), but the peaks do not look quite like a λ; the heat capacity peak [3.24] for the PyAg$_5$I$_6$ transition at about 320 K is similar to these. MgF$_2$ does not have the fluorite structure, does not show this transition and presumably is not a solid electrolyte; K$_2$S has the antifluorite structure and must be a solid electrolyte with cation (K$^+$ ion) current carriers [3.95]. *Dworkin* and *Bredig* [3.95] also point out the analogy of the highly conducting high temperature Li$_2$SO$_4$ phase [3.96] to the antifluorite K$_2$S.

As in the case of PyAg$_5$I$_6$ (see Sect. 3.1.1) the σ vs $1/T$ plot for CaF$_2$ shows a noticeable change of slope at the transition [3.84]. The observations suggest that in the fluorite and antifluorite structures the transition indicated by the heat capacity and conductivity data is from a crystal structure of low disorder to one

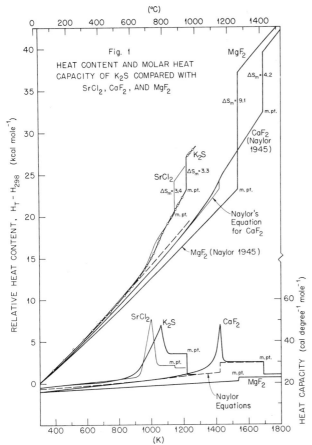

Fig. 3.8. Heat contents and heat capacities of CaF_2, $SrCl_2$, K_2S and MgF_2. (Reprinted with permission from *A. S. Dworkin* and *M. A. Bredig*. J. Am. Chem. Soc. **72**, 1277 (1968). Copyright by the American Chemical Society)

of high disorder. The structure changes continuously from well below the transition to the melting point, even though the *average* space group does not change. Below the transition not all the current carriers are mobile, whereas above the transition it is probable that all are mobile (see also [3.95, 97]).

Nagel and *O'Keeffe* [3.85] have shown (see also [3.83]) that $Ca_{0.75}Y_{0.25}F_{2.25}$ is a high temperature solid electrolyte with higher conductivity than CaF_2, SrF_2, and BaF_2. One might be led to think that this implies a cation vacancy structure, but according to *Cheetham* et al. [3.98] it does not. From a neutron diffraction investigation they find two types of interstitial sites within the cubes. In $Ca_{0.75}Y_{0.25}F_{2.25}$, these are at $(0.5, 0.381, 0.381) + fcc(F')$ and $(0.359, 0.359, 0.359) + fcc(F'')$. Per formula unit, there are 1.62 F^- ions in normal sites, 0.47 in F' sites and 0.16 in F'' sites. The authors show possible models involving clustering of defects that would account for these occupancies.

3.4 Other Halogenide Solid Electrolytes

Nagel and *O'Keeffe* [3.85] and *O'Keeffe* [3.99] have shown that a number of compounds and solid solutions with LaF_3-related structures behave like solid electrolytes. The crystal structure of LaF_3 has been reported by *Mansmann* [3.100] and by *Zalkin* et al. [3.101]. (It has recently been contested by *Afansiev* et al. [3.102].) *O'Keeffe* and *Hyde* [3.103] have demonstrated a relationship between the LaF_3 and YF_3 structures and that of the high temperature Cu_2S [3.104] structure and thereby indicated the probable pathways for motion of the fluoride ions.

Krogh-Moe et al. [3.105] have shown that K_2BaCl_4 undergoes an abrupt and large increase in conductivity at 635° C. The material is probably an anion conductor [3.106]. The crystal structure of this material both below and above the transition would be of considerable interest.

3.5 Final Note

An excellent review [3.107] of AgI-based solid electrolytes appeared before the manuscript went to the editor of this series. Consequently, a subsection on results of the now numerous investigations on α-AgI was not included in this chapter. This particular subject has been covered in much greater detail in the article by *Funke* [3.107] than is possible here. Some overlapping between Section 3.1 and Funke's article remains. Nevertheless, in the larger part, the articles are different in both emphasis and content.

Acknowledgement. The author's research in the field of solid electrolytes is supported by the National Science Foundation.

References

3.1   S.Geller: Science **157**, 310 (1967)
3.2   L.D.Fullmer, M.A.Hiller: J. Cryst. Growth **5**, 395 (1969)
3.3   B.B.Owens: Solid Electrolytie Batteries. In *Advances in Electrochemistry and Electrochemical Engineering*, Vol. 8, ed. by C. W. Tobias (John Wiley and Sons, Inc., New York 1971) pp. 1–62
3.4   D.O.Raleigh: J. Appl. Phys. **41**, 1876 (1970)
3.5   B.B.Owens, G.R.Argue: J. Electrochem. Soc. **117**, 898 (1970)
3.6   J.N.Bradley, P.D.Greene: Trans. Faraday Soc. **63**, 424 (1967)
3.7   B.B.Owens, G.R.Argue: Science **157**, 308 (1967)
3.8   J.N.Bradley, P.D.Greene: Trans. Faraday Soc. **62**, 2069 (1966)
3.9   S.Geller: Unpublished
3.10 F.Bonnini, M.Lazzari: J. Power Sources **1**, 103 (1976)
3.11 B.Scrosati, G.Pistoia, M.Lazzari, L.Peraldo Bicelli: J. Appl. Electrochem. **4**, 201 (1974)
3.12 S.Geller, B.B.Owens: Unpublished
3.13 W.V.Johnston, H.Wiedersich, G.W.Lindberg: J. Chem. Phys. **51**, 3739 (1969)
3.14 F.L.Lederman, M.B.Salamon, H.Peisl: Solid State Commun. **19**, 147 (1976)
3.15 S.Geller: Phys. Rev. **B 14**, 4345 (1976)
3.16 L.J.Graham, R.Chang: J. Appl. Phys. **46**, 2433 (1975)

3.17 H. Wiedersich, S. Geller: Properties of Highly Conducting Halide and Chalcogenide Solid Electrolytes. In *The Chemistry of Extended Defects in Non-Metallic Solids*, ed. by L. Eyring, M. O'Keeffe

3.18 W. J. Pardee, G. D. Mahan: J. Solid State Chem. **15**, 310 (1975)

3.19 B. B. Owens: J. Electrochem. Soc. **117**, 1536 (1970)

3.20 B. B. Owens: Thermodynamic Properties of Solid Electrolytes. In *Fast Ion Transport in Solids*, ed. by W. van Gool (North Holland, Amsterdam-London 1973) pp. 593–606

3.21 S. Geller, M. D. Lind: J. Chem. Phys. **52**, 5854 (1970)

3.22 S. Geller, B. B. Owens: J. Phys. Chem. Solids **33**, 1241 (1972)

3.23 S. Geller: Science **176**, 1016 (1972)

3.24 T. Hibma: Phys. Rev. B**15**, 5797 (1977)

3.25 T. Hibma, S. Geller: J. Solid State Chem. **21**, 225 (1977)

3.26 H. Hartl: Acta Cryst. B**31**, 1781 (1975)

3.27 *International Tables for X-Ray Crystallography*, ed. by N. F. Henry, K. Lonsdale (Kynoch Press, Birmingham 1969)

3.28 L. D. Landau, E. M. Lifschitz: *Statistical Physics* (Addison-Wesley, Reading, Mass. 1969)

3.29 T. Kaneda, E. Mizuki: Phys. Rev. Letters **29**, 937 (1972)

3.30 S. Geller, P. M. Skarstad: Phys. Rev. Letters **33**, 1484 (1974)

3.31 S. Geller, P. M. Skarstad, S. A. Wilber: J. Electrochem. Soc. **122**, 332 (1975)

3.32 L. Y. Y. Chan, S. Geller: J. Solid State Chem. **21**, 331 (1977)

3.33 T. Takahashi, S. Ikeda, O. Yamamoto: J. Electrochem. Soc. **120**, 647 (1973)

3.34 P. M. Skarstad, S. Geller: Mat. Res. Bull. **10**, 791 (1975)

3.35 C. Tubandt, E. Lorenz: Z. Physik. Chem. **87**, 513 (1914)

3.36 L. W. Strock: Z. Physik. Chem. B **25**, 441 (1934)

3.37 L. W. Strock: Z. Physik. Chem. B **31**, 132 (1936)

3.38 W. Bührer, W. Hälg: Helv. Phys. Acta **47**, 27 (1974)

3.39 R. M. Cava, B. J. Wuensch: The Distribution of Silver Ions in α-AgI. In *Superionic Conductors*, ed. by G. D. Mahan, W. L. Roth (Plenum Press, New York and London 1976) pp. 217–218

3.40 P. Rahlfs: Z. Physik. Chem. B **31**, 157 (1936)

3.41 S. Hoshino: J. Phys. Soc. Japan **7**, 560 (1952)

3.42 B. Reuter, K. Hardel: Naturwissenschaften **48**, 161 (1961)

3.43 B. Reuter, K. Hardel: Z. Anorg. Allg. Chem. **340**, 158 and 168 (1965)

3.44 B. Reuter, K. Hardel: Ber. Bunsenges. Phys. Chem. **70**, 82 (1966)

3.45 T. Takahashi, O. Yamamoto: Electrochim. Acta **11**, 779 (1966)

3.46 T. Takahashi, O. Yamamoto: Electrochim. Acta **11**, 911 (1966)

3.47 T. Takahashi: J. Appl. Electrochem. **3**, 79 (1973)

3.48 J. A. A. Ketelaar: Z. Krist. (A) **87**, 436 (1934)

3.49 J. A. A. Ketelaar: Trans. Faraday Soc. **34**, 874 (1938)

3.50 S. Miyake, S. Hoshino, T. Takenaka: J. Phys. Soc. Japan **7**, 19 (1952)

3.51 J. S. Kasper, K. W. Browall: J. Solid State Chem. **13**, 49 (1975)

3.52 K. W. Browall, J. S. Kasper: J. Solid State Chem. **15**, 54 (1975)

3.53 J. A. A. Ketelaar: Z. Phys. Chem. B **26**, 327 (1934)

3.54 T. Hibma, H. U. Beyeler, H. R. Zeller: J. Phys. C: Solid State Phys. **9**, 1691 (1976)

3.55 B. B. Owens, J. H. Christie, G. T. Tiedeman: J. Electrochem. Soc. **118**, 1144 (1971)

3.56 J. H. Christie, B. B. Owens, G. T. Tiedeman: to be published

3.57 J. H. Christie, B. B. Owens, G. T. Tiedeman: Abstracts, 146th Meeting of the Electrochem. Soc., New York, 1974, No. 20

3.58 M. L. Berardelli, C. Biondi, M. DeRossi, G. Fonseca, M. Giomini: J. Electrochem. Soc. **119**, 114 (1973)

3.59 T. Takahashi, S. Ikeda, O. Yamamoto: J. Electrochem. Soc. **119**, 477 (1972)

3.60 T. Takahashi, E. Nomura, O. Yamamoto: J. Appl. Electrochem. **2**, 51 (1972)

3.61 T. Takahashi, K. Kuwabara, O. Yamamoto: J. Electrochem. Soc. **120**, 1607 (1973)

3.62 T. Takahashi, O. Yamamoto: J. Electrochem. Soc. Japan **34**, 211 (1966)

3.63 L. Suchow, G. R. Pond: J. Phys. Chem. **58**, 240 (1954)

3.64   J.Coetzer, M.M.Thackeray: Electrochim. Acta **21**, 37 (1976)
3.65   J.H.Christie, B.B.Owens, G.T.Tiedeman: Inorg. Chem. **14**, 1423 (1975)
3.66   S.Geller: Crystal Structure and Conductivity in AgI-based Solid Electrolytes. In *Fast Ion Transport in Solids*, ed. by W. van Gool (North Holland, Amsterdam, London 1973) pp. 607–616
3.67   S.Geller: Crystal Structure and Conductivity of AgI-based Solid Electrolytes II. In *Superionic Conductors*, ed. by G. D. Mahan, W. L. Roth (Plenum Press, New York, London 1976) pp. 171–181
3.68   H.Wiedersich, W.V.Johnston: J. Phys. Chem. Solids **30**, 475 (1969)
3.69   R.D.Armstrong, R.S.Bulmer, T.Dickinson: J. Solid State Chem. **8**, 219 (1973)
3.70   L.Heyne: Electrochim. Acta **15**, 1251 (1970)
3.71   D.Emin, T.Holstein: Ann. Phys. (New York) **53**, 439 (1969)
3.72   G.G.Bentle: J. Appl. Phys. **39**, 4036 (1968)
3.73   M.Nagao, T.Kaneda: Phys. Rev. B **11**, 2711 (1975)
3.74   K.E.Johnson, S.J.Sime, J.Dudley: J. Electrochem. Soc. **120**, 703 (1973)
3.75   A.V.Danilov, V.E.Ivanov, S.V.Karpov: Soviet Phys. Solid State **16**, 1259 (1975)
3.76   S.Chandra, H.B.Lal, K.Shahi: J. Phys. D. **5**, 443 (1972)
3.77   R.S.Bauer, B.A.Huberman: Phys. Rev. B **13**, 3344 (1976)
3.78   G.Burns, F.H.Dacol, M.W.Shafer: Solid State Commun. **19**, 287 (1976)
3.79   M.J.Delaney, S.Ushioda: Solid State Commun. **19**, 297 (1976)
3.80   G.Burns, F.H.Dacol, M.W.Shafer: Solid State Commun. **19**, 291 (1976)
3.81   T.Takahashi, O.Yamamoto, S.Ikeda: J. Electrochem. Soc. **120**, 1431 (1973)
3.82   A.F.Sammells, J.Z.Gougoutas, B.B.Owens: J. Electrochem. Soc. **122**, 1291 (1975)
3.83   R.W.Ure,Jr.: J. Chem. Phys. **26**, 1363 (1957)
3.84   C.E.Derrington, A.Lindner, M.O'Keeffe: J. Solid State Chem. **15**, 171 (1975)
3.85   L.E.Nagel, M.O'Keeffe: Highly Conducting Fluorides Related to Fluorite and Tysonite. In *Fast Ion Transport in Solids*, ed. by W. van Gool (North Holland, Amsterdam, London 1973) pp. 165–172
3.86   C.E.Derrington, M.O'Keeffe: Solid State Commun. **15**, 1175 (1974)
3.87   J.Schoonman, G.J.Dirksen, G.Blasse: J. Solid State Chem. **7**, 245 (1973)
3.88   C.E.Derrington, M.O'Keeffe: Nature Phys. Sci. **246**, 44 (1973)
3.89   S.M.Shapiro: Neutron Scattering Studies of Solid Electrolytes. In *Superionic Conductors*, ed. by G. D. Mahan, W. L. Roth (Plenum Press, New York, London 1974) pp. 261–277
3.90   K.Funke, R.Hackenburg: Ber. Bunsenges. Phys. Chem. **76**, 885 (1972)
3.91   G.Eckold, K.Funke: Z. Naturforsch. **28**a, 1042 (1973)
3.92   G.Eckold, K.Funke, J.Kalus, R.E.Lechner: Phys. Lett. **55** A, 125 (1975)
3.93   C.Clemon, K.Funke: Ber. Bunsenges. Phys. Chem. **79**, 1119 (1975)
3.94   B.F.Naylor: J. Am. Chem. Soc. **67**, 150 (1945)
3.95   A.S.Dworkin, M.A.Bredig: J. Am. Chem. Soc. **72**, 1277 (1968)
3.96   A.Kvist, A.Lunden: Z. Naturforsch. **20**a, 235 (1965)
3.97   M.A.Bredig: The Order-Disorder (Lambda) Transition in Uranium Dioxide and Other Solids of the Fluorite Type Structure. In *Colloques internationaux CNRS, No.* 205 *Études des Transformations Cristallines à Haute Température au-dessus de* 2000 K. Sept. 27–30, 1971
3.98   A.E.Cheetham, B.E.F.Fender, M.J.Cooper: J. Phys. C **4**, 3107 (1971)
3.99   M.O'Keeffe: Science **180**, 1276 (1973)
3.100 M.Mansmann: Z. Krist. **122**, 375 (1965)
3.101 A.Zalkin, D.H.Templeton, T.E.Hopkins: Inorg. Chem. **5**, 1466 (1966)
3.102 M.L.Afansiev, S.P.Habuda, A.G.Lundin: Acta Cryst. B **28**, 2903 (1972)
3.103 M.O'Keeffe, B.G.Hyde: J. Solid State Chem. **13**, 172 (1975)
3.104 M.J.Buerger, B.J.Wuensch: Science **141**, 276 (1963)
3.105 J.Krogh-Moe, M.Vikan, C.Krohn: Acta Chem. Scand. **21**, 309 (1967)
3.106 H.Schmalzried: Z. Physik. Chem. N.F. **33**, 129 (1962)
3.107 K.Funke: Progr. in Solid State Chem. **11**, 345 (1976)

4. Applications of Halogenide Solid Electrolytes

B. B. Owens, J. E. Oxley, and A. F. Sammells

With 17 Figures

In this chapter the authors briefly discuss the device related characteristics of halogenide electrolytes. Following that will be a discussion of the general principles involved in the design of solid state electrochemical cells, and then in later sections various applications of solid electrolytes to practical devices will be described. The applications described in this chapter include devices that are commercially available as well as devices that are still in the conceptual stage.

4.1 Materials

Several classes of halogenide solid electrolytes have been investigated for device application. The following four groups are considered here: silver ion conductors, copper ion conductors, lithium ion conductors and halide ion conductors.

4.1.1 Silver Ion Conducting Solid Electrolytes

This category comprises the largest number of known solid state electrolytes and includes those which have the highest low temperature ionic conductivity. The group is exemplified by the compound $RbAg_4I_5$ which has a conductivity of 0.27 $[\Omega\,cm]^{-1}$ at 25° C. Properties of this class of materials are described in Section 3.1 as well as in some earlier reviews [4.1, 2]. The key characteristics for device applications are the current carrying species and the temperature range of stability of the conductive phase. In many respects $RbAg_4I_5$ behaves as an ideal solid electrolyte in that the Ag^+ ion transport number is close to unity, and the electronic transport number is less than 10^{-10}. Similar transport properties are exhibited by many of the other Ag^+ ion solid electrolytes, but none is more conductive than $RbAg_4I_5$. On the negative side, $RbAg_4I_5$ exhibits the low decomposition voltage of 0.66 V and is thermodynamically unstable below 27° C [4.3].

4.1.2 Copper Ion Conducting Solid Electrolytes

Copper ion conduction in halogenide double salts has been reported in a number of compounds as described in Section 3.2. Some limited device development has been reported on these materials, but they appear to suffer from an inherent instability, and no practical applications have been developed up to the present time [4.4].

4.1.3 Lithium Ion Conducting Solid Electrolytes

Table 4.1 compares the conductivity of the lithium halides at 25° C. Lithium iodide is the most conductive and has been utilized in solid state batteries operative at room temperature.

First approaches towards increasing the ionic conductivity of LiI involved the substitution of divalent cations. The increase in conductivity resulting from the substitution for a small percentage of lithium ions has been explained by *Schlaikjer* and *Liang* [4.7] as due to an increase in the number of Schottky defects leading to an increase in the number of cation vacancies available for the conduction process. The doping of LiI by Ca^{2+} at a concentration[1] of 1 m/o CaI_2 was reported to give material with an initial conductivity of $1.2 \cdot 10^{-5}$ $[\Omega cm]^{-1}$ immediately after preparation at 25° C; however, it was observed that deterioration of the conductivity occurred. After about three weeks the conductivity equilibrated at about $2 \cdot 10^{-6}$ $[\Omega cm]^{-1}$. Therefore, any device based on this electrolyte is restricted to a high internal resistance.

It was subsequently reported by *Liang* that the ionic conductivity of LiI could be increased not only by the introduction of dopants, but also by the preparation of solid state mixtures of Al_2O_3 and LiI [4.1, 8]. Stable maximum conductivities of 10^{-5} $[\Omega cm]^{-1}$ were achieved with such solid state mixtures, this being more than one order of magnitude higher than the stable conductivities previously achieved with conventionally doped LiI. The maximum conductivity of the LiI–Al_2O_3 electrolyte occurred with material containing 30–35 m/o Al_2O_3. No significant electronic conductivity could be detected in these materials as was shown by the discharge characteristics of Li–PbI_2 cells stored for one year at 25°, 40°, and 60° C [4.9]. Subsequently, *Liang* reported that there was no loss in capacity following two years of storage under uncontrolled ambient conditions [4.10].

Recently, *Owens* and *Hanson* [4.11] discovered that the conductivity of LiI could be increased to about 10^{-5} $[\Omega cm]^{-1}$ by the addition of a controlled amount of water sufficient to give the monohydrate species. To aid in the incorporation of water in this electrolyte and to improve the handling properties of the solid electrolyte high surface area silica or alumina was added to the electrolyte. In this electrolyte it appears that $LiI \cdot H_2O$ coexists with LiI and the SiO_2 or Al_2O_3. Further investigation demonstrated that the pure monohydrate has a specific conductivity of 10^{-5} $[\Omega cm]^{-1}$. Transference number experiments to determine whether protons were involved in the conduction process were not performed. However, electrochemical cells, Li/$LiI \cdot H_2O$, LiI, SiO_2/I_2, were shown to have normal voltages (2.75 V for I_2 complexed with Me_4NI) and to

---

[1]   Compositions will be expressed as mole percent (m/o) or weight percent (w/o).

Table 4.1. Nominal values of specific conductivity of lithium halides at 25 °C

Salt	Specific conductivity $[\Omega \, cm]^{-1}$
LiF[a]	10^{-13}
LiCl[b]	$5 \cdot 10^{-10}$
LiBr[b]	$2 \cdot 10^{-9}$
LiI[b]	$1 \cdot 10^{-7}$

[a] Extrapolation of data [4.5].
[b] [4.6].

Table 4.2. Conductivity of LiX-H_2O-metal oxide solid electrolytes at 25 °C [4.11]

Electrolyte (empirical composition)	$[\Omega \, cm]^{-1}$
LiI + 25 w/o Al_2O_3 + 11 w/o H_2O	$4 \cdot 10^{-5}$
LiI + 6 w/o SiO_2 + 10 w/o H_2O	$1 \cdot 10^{-5}$
LiBr + 34 w/o Al_2O_3 + 8 w/o H_2O	$2 \cdot 10^{-6}$
LiBr + 22 w/o SiO_2 + 14 w/o H_2O	$1 \cdot 10^{-5}$
LiCl + 35 w/o Al_2O_3 + 19 w/o H_2O	$1 \cdot 10^{-4}$
LiCl + 22 w/o SiO_2 + 17 w/o H_2O	$8 \cdot 10^{-6}$
LiF + 34 w/o Al_2O_3 + 10 W/o H_2O	$1 \cdot 10^{-4}$

deliver high capacities, with formation of LiI product in the cathode. Further, mobile protons would be reduced at the lithium surface, evolving H_2. Therefore, it was concluded that Li^+ ions are the dominant current carrying species with no significant contributions from electronic conduction or other ionic species.

Owens and *Hanson* [4.11] also investigated the other lithium halides in combination with H_2O and metal oxides. The results are summarized in Table 4.2. The x-ray analyses of the solid electrolytes indicated the presence of the monohydrate phase for all except LiF. Further investigation is needed on these materials. However, preliminary indications are that they are all solid Li^+ ion conductors that can be used in solid state electrochemical devices.

The decomposition voltage of these electrolytes is governed by the electronegative halide in the mixture. LiI based electrolytes are limited to 2.8 V, whereas the decomposition voltages are increased to nominal values of 3.5, 4, and 6 V for LiBr, LiCl, and LiF, respectively. The significance of this series of electrolytes with regard to device application is that they permit cells with voltages and energies significantly higher than those of LiI based materials.

4.1.4 Halide Ion Conductors

Halide ion conductivity has been found to date in a wide variety of materials (see Secs. 3.3 and 3.4). One of the more recently investigated halide ion conductors, lead fluoride, has been utilized in electrochemicals cells. Lead fluoride has been observed to exist in two crystallographic forms, α-PbF$_2$ which has orthorhombic symmetry (density 8.445) and is the most commonly found form of lead fluoride and β-PbF$_2$ which has been observed to have cubic symmetry (density 7.750) [4.12]. Evaluation of lead fluoride as a solid electrolyte was initiated several years ago [4.13]; however, with this work it was not apparent which of the crystallographic variations of lead fluoride was in fact being investigated. The material used may well have been a mixture of both. The temperature at which conversion of α-PbF$_2$ to β-PbF$_2$ occurs has been studied using DTA techniques [4.14] and a transition temperature in the range of 334–344° C has been observed, the exact temperature being somewhat dependent upon the heating rate used. Conversion of β-PbF$_2$ to α-PbF$_2$ can be effected by pressures above 30 000 psi.

Kennedy et al. [4.14] determined the relative stability between these two phases by use of the cell Pb/β-PbF$_2$//KF(aq)//α-PbF$_2$/Pb. At equilibrium the open circuit potential for this cell was $+1.7$ mV with the probable half cell reactions

$$Pb + 2F^- \rightarrow \beta\text{-PbF}_2 + 2e \tag{4.1}$$

$$\alpha\text{-PbF}_2 + 2e \rightarrow Pb + 2F^- \tag{4.2}$$

giving the overall reaction

$$\alpha\text{-PbF}_2 \rightarrow \beta\text{-PbF}_2. \tag{4.3}$$

The EMF of this cell represents the free energy difference between the two forms of lead fluoride and corresponds to -78 cal/mole, indicating greater stability for the β-PbF$_2$ at room temperature. Conductivity values achieved from pressed pellets at about 150° C have been reported at $6 \cdot 10^{-5}$ $[\Omega\,\text{cm}]^{-1}$ for β-PbF$_2$ and around $5 \cdot 10^{-6}$ $[\Omega\,\text{cm}]^{-1}$ for α-PbF$_2$ [4.14]. Conductivity measurements made on single crystal samples of β-PbF$_2$ gave conductivities of $2 \cdot 10^{-8}$ $[\Omega\,\text{cm}]^{-1}$ at 25° C and $6 \cdot 10^{-5}$ $[\Omega\,\text{cm}]^{-1}$ at 150° C. The activation energy observed with such single crystals was 13.8 kcal/mole as compared with 11.5 kcal/mole for polycrystalline PbF$_2$. This larger activation energy for single crystal material may indicate the presence of possible grain boundary conduction in the polycrystalline samples. The electronic conductivity of β-PbF$_2$ was determined with a cell of configuration Pb/β-PbF$_2$/Au. The lead and gold electrodes were vacuum deposited onto β-PbF$_2$ pressed pellets, and electronic conductivities of about 10^{-11} $[\Omega\,\text{cm}]^{-1}$ at 159° C were observed, representing only 10^{-4}% of the total conductivity at this temperature. The transport number for fluoride ions was determined by use of the cell

Cell I   Pb/β-PbF$_2$//β-PbF$_2$//β-PbF$_2$/Au

which consists of three β-PbF$_2$ pellets held together under pressure. Based on the relative weight changes in the pellets on the right-hand and left-hand side of the cell, the transport number for F$^-$ was 0.93 and for Pb^{2+} was 0.07.

Recently, *Schoonman* et al. [4.15] reported the ionic conductivity of BiO_xF_{3-2x} ($x=0.09-0.1$) at $25°C$ to be 10^{-6} $[\Omega\,cm]^{-1}$. The use of this electrolyte in electrochemical cells has also been recently described [4.16].

4.2 Electrochemical Devices

Electrochemical devices differ from electronic devices in that the passage of current requires the motion and transfer of both ions and electrons. They are thus not passive in a mechanical sense and whenever an electrochemical device is incorporated into an electronic circuit, it will inevitably emerge as the weakest link. This is particularly true for multicomponent liquid electrolyte systems because of the larger number of charge transfer or faradaic reactions which are possible. In a system based on an aqueous electrolyte one must thus take into account not only reactions associated with an active oxidant or reductant, but also the numerous electrode reactions of the H_3O^+ or OH^- ions. The practicality of any electrochemical device concept which relies on the use of an aqueous electrolyte is often, therefore, a compromise between the needs for acceptable materials for long term stability and the obtaining of satisfactory performance.

The ideal electrolyte would be both chemically and physically invariant. It should at the same time provide a path for ionic transport, a barrier to electronic transport, and serve as a mechanical separator for the electrodes. Solid electrolytes go a long way towards fulfilling these requirements. However, the number of materials having sufficient low temperature ionic conductivity to permit their use in practical devices is largely restricted to the Ag^+ ion conducting families and to a lesser extent some Li^+ ion conductors. The widest range of devices demonstrated to date has been based on Ag^+ ion electrolytes and in particular on $RbAg_4I_5$. The properties of $RbAg_4I_5$ are summarized in Table 4.3. The greatest commercial interest today, however, is in Li^+ ion conductors which are beginning to find wide application in high energy density

Table 4.3. Physical and chemical constants of $RbAg_4I_5$ [4.17]

Parameter	Constant
Diffusion coefficient of Ag^+	$D_0 = 3\cdot10^{-6}$ cm^2/s
Transport number of Ag^+	$t_{Ag^+} = 1.00 \pm 0.01$
Density	$d = 5.38$ gm/cm^3
Standard free energy of formation	$\Delta G^0 = -142.3 \pm 0.1$ kcal/mole
Electronic conductivity	$\sigma < 10^{-11}$ $[\Omega\,cm]^{-1}$
Ionic conductivity	$+75\,°C, \sigma = 0.39$ $[\Omega\,cm]^{-1}$
	$+25\,°C, \sigma = 0.26$ $[\Omega\,cm]^{-1}$
	$-55\,°C, \sigma = 0.09$ $[\Omega\,cm]^{-1}$
Melting point	$232\,°C$
Decomposition potential	0.66 V

primary batteries for use in specialized applications such as cardiac pacemakers. The larger number of devices developed for the Ag^+ ion conductors, while offering some unique potentialities based on their very low ionic resistivity, have as yet failed to develop substantial commercial importance.

4.2.1 Electrode Reactions

The choice of electrode materials for solid state electrochemical cells is governed by the fact that, in general, all the current is carried by a single ion. Thus in $RbAg_4I_5$, $t_{Ag^+} = 1$, and silver metal is the only choice for the negative electrode (anode) in a galvanic cell. At the positive (cathode), the choice is wider but two criteria must be fulfilled: the cathode reversible potential must be less positive than the decomposition potential of the electrolyte and the half-cell reaction must be electrically compatible with the mobile ion.

In practical galvanic cells based on $RbAg_4I_5$, both electrode structures are fabricated with the electroactive materials in finely divided form closely admixed with electrolyte and carbon. The latter serves both as current collector material and as a means of maximizing the internal surface area. The presence of electrolyte in the electrode structure is, of course, necessary to maximize the electrode/electrolyte interface area but is also essential as a binder to reinforce the bulk interface contact between the electrode and electrolyte components. Carbon is electrochemically inert in this system and is ideally polarizable over the potential range ~ -20 to $+500\,mV$ vs Ag/Ag^+. The current which flows in this region is thus purely capacitive. The fact that the carbon can be polarized negative to the silver potential is a consequence of the existence of a nucleation overpotential for the deposition of silver onto carbon [4.18]. At potentials above $\sim 500\,mV$, the faradaic reaction associated with electrolyte decomposition is dominant and the carbon ceases to behave purely capacitively.

It is essential that impurities be excluded from these systems. Both H_2O and mixtures of O_2 and H_2O were shown [4.19] to give impurity currents. These impurity currents masked early attempts to determine the level of electronic conductivity in electrolytes such as $RbAg_4I_5$ using the Wagner polarization technique. This was simply shown in a blocking electrode experiment to determine electronic conductivity using the cell,

Cell II $Ag/RbAg_4I_5/C$ (blocking electrode).

Comparative measurements were made using both planar carbon pressed against the electrolyte layer and a mixture of carbon and $RbAg_4I_5$, formed into a pellet and pressed against the electrolyte layer. The currents measured when the blocking electrode was held potentiostatically at a point equidistant between the silver and iodine potentials were found to be several orders of magnitude greater in the case of the mixed carbon/$RbAg_4I_5$ electrode. Furthermore, the number of coulombs passed was greatly in excess of that required to establish the equilibrium activity of iodine at this potential. It was further observed that by

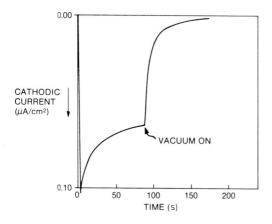

Fig. 4.1. Effect of vacuum on impurity currents

placing the cell in a vacuum chamber the impurity currents were substantially reduced, see Fig. 4.1. All subsequent measurements of electronic conductivity on Ag^+ ion conducting electrolytes were thus made in a vacuum chamber, and a value of electronic conductivity was not assigned to any material until the current had stabilized. This was found to take up to several weeks at 10^{-10} Torr but the values ultimately measured probably still only represent the upper limit of electronic conductivity.

Another more practical implication of the role of impurities is in their effect on electrode performance and stability [4.19]. At the silver potential, the net impurity reaction in the presence of O_2 and H_2O vapor was observed to be cathodic and presumably corresponded to the reaction

$$O_2 + 2H_2O + 4e \rightarrow 4OH^-. \tag{4.4}$$

This was directly confirmed by chemical analysis of the degradation of mixtures of $Ag/RbAg_4I_5/C$ in the presence of O_2 and H_2O vapor. The overall reaction found was

$$O_2 + 2H_2O + 4Ag + 4RbAg_4I_5 \rightarrow 4RbOH + 20AgI. \tag{4.5}$$

It is thus important to avoid the presence of atmospheric impurities both during materials preparation and cell assembly as well as to ensure that the final cell is adequately sealed to prevent ingress during the useful life of the cell. A hermetic seal is desirable and it is, of course, essential that all cell assembly be carried out in an inert atmosphere dry box. This is in any case a requirement for devices employing $RbAg_4I_5$ because of the role of moisture in causing the disproportionation of this material at temperatures $<27\,°C$ [4.3].

A further implication of impurity reactions was found with silver anodes. To achieve significant utilization of the silver beyond the first monolayer exposed to the electrolyte, it is necessary to postulate movement of silver metal, probably via surface self-diffusion, as the silver is progressively oxidized. It was found that almost complete electrochemical utilization ($>95\%$) of the silver was achieved in anodes which were kept free of moisture, even at low temperatures (e.g., $-55\,°C$), whereas even partial exposure to moisture vapor reduced this utilization to $<10\%$ at low temperature. It is postulated that adsorption of the impurity species on the clean silver surface inhibits its surface self-diffusion.

The requirement for an ideal solid state cell system is that both of the electrodes be solid and capable of being charged and discharged at acceptable current densities. Although metal electrodes, silver in the case of Ag^+ ion conductors and lithium in the case of Li^+ ion conductors, fulfill this requirement, an efficient high capacity solid state positive electrode is more difficult to achieve. One is restricted in the choice to those couples which involve participation of the conducting ion, e.g., Ag^+ or Li^+. An exception to this, as will be discussed later, is in applications needing only very low coulombic capacity. Here one can avoid the use of an active oxidant and instead rely on the charge storage capability of a high surface area carbon/solid electrolyte interface. This can, as indicated, be purely capacitive or, making use of the undervoltage electrolyte decomposition, involve adsorption of up to a monolayer of (e.g.) iodine at a carbon or other inert electrode/electrolyte interface.

Although some degree of success has been reported using cathodes containing an active solid state oxidant, *vide infra*, most of the practical battery systems developed to date have relied on the use of iodine, albeit complexed and thus having reduced activity, as the active oxidant. These cathodes thus contain, depending on the degree of complexation, free iodine and have a finite vapor pressure. Low electrode polarization (concentration) is achieved by virtue of the ability of the iodine to diffuse within the electrode structure. It is recognized that the use of "free" iodine compromises the whole concept of a solid state battery, and several precautions must be taken; first, to contain it within the electrode structure and to minimize its diffusion across the electrolyte to the negative electrode, and secondly, to protect the cell against corrosion.

The ability to recharge a solid state electrochemical cell is limited by several factors. Consider first the negative electrode, e.g., Ag or Li. Irrespective of the starting capacity density (Ah/cm^2) of an electrode structure, the capacity which can be returned is limited because the metal will not be replated inside the porous structure so long as the limiting diffusion current for metal deposition is not exceeded. The latter is governed by the product of the ionic mobility and charge carrier concentration, which for the highly conducting silver ion electrolytes is several A/cm^2. At current densities lower than this, ohmic factors will predominate and the metal will be plated out only at the electrode surface nearest to the positive electrode. Mechanical factors associated with stress caused by growth of the electrodeposit together with the inherent tendency for metals to electrodeposit in crystalline form, i.e., dendritically, will thus limit the charge acceptance. Although dendrites are not able to grow through a single ionic crystal, they are nevertheless expected to propagate and expand along stress-induced voids and crevices in a polycrystalline matrix.

Recharging of the positive electrode in a solid state cell is more complex. Consider first Cell II: at positive potentials, iodide ions are oxidized to form free iodine which is adsorbed at the carbon/electrolyte interface. If a voltage is imposed on the cell in what may be termed the "pseudocapacity" region, $0.5\,V < E < 0.66\,V$, the current will decay exponentially while the required thermodynamic activity of iodine is established. Unit activity corresponding to approximately monolayer coverage is attained at 0.66 V, the electrolyte decomposition voltage. Above 0.66 V, the electrolyte will decompose freely to a limit determined by the availability of iodide ions at the carbon/electrolyte interface and "removal" of the iodine which is generated. In the case of $RbAg_4I_5$ some of the iodine will go to form the compound RbI_3, either by decomposition of $RbAg_4I_5$ or by equilibration with RbI. The steps in the overall electrolyte decomposition reaction can be set down as follows:

$$RbAg_4I_5 \rightarrow 4Ag + 2I_2 + RbI \qquad (4.6)$$

$$I_2 + RbAg_4I_5 \rightarrow RbI_3 \qquad (4.7)$$

and the overall reaction is

$$2RbAg_4I_5 \rightarrow 4Ag + 2RbI_3 + 4AgI \qquad (4.8)$$

with the concurrent equilibration,

$$RbI_3 \rightleftharpoons RbI + I_2 . \qquad (4.9)$$

Charge acceptance is poor as a result of the buildup of a high internal resistance. The following possibilities exist for the buildup of this resistance.
  1) Formation of poorly conducting RbI_3 and/or AgI.
  2) Congregation of poorly conducting iodine, $\sigma = 10^{-9}\,[\Omega\,cm]^{-1}$ (electronic), at the carbon/electrolyte interface.
  3) Development of poor contact at the carbon/electrolyte interface due to immobility of the iodide ions.
  In contrast to the silver deposition process, iodine can be formed inside the electrode structure, e.g., a carbon/$RbAg_4I_5$ composite, because the iodide ions are not intrinsically mobile and limiting concentration polarization is readily established. In primary cells already containing a source of iodine, e.g., as a polyiodide, the same phenomena apparently govern and resistance builds up in these systems also.

4.2.2 Device Concepts

Four main device concepts utilizing solid electrolytes will be discussed in this chapter, namely, primary batteries, capacitors, rechargeable batteries and coulometers. Primary attention will be devoted to those device concepts which have developed around the family of Ag^+ ion conductors typified by $RbAg_4I_5$. In the case of primary batteries, however, a discussion is included also of lithium batteries. Specific examples and descriptions will be provided of the various types of devices which have been developed.

4.3 Primary Batteries

There is a considerable body of literature describing solid electrolyte cells. Because the present chapter is primarily concerned with applications, the authors will restrict this review to the practical systems developed for operation at normal ambient temperatures. The systems to be described are the following:

Cell III	$Ag/RbAg_4I_5/I_2$
Cell IV	$Li/LiI/PbI_2, PbS$
Cell V	$Li/LiI/I_2$
Cell VI	$Li/LiBr/Br_2$.

4.3.1 Silver-Iodine Solid State Batteries

There are many solid electrolyte cells based on Ag^+ ion conducting electrolytes, see, for example, earlier review articles by *Liang* [4.1] and *Owens* [4.2]. More recent reports of new solid Ag^+ ion conducting electrolytes have resulted in continuing investigations of new cell systems. However, all these cells are intrinsically limited to the use of silver anodes and cathodes that are thermodynamically compatible with the electrolyte. Thus, all these cells are limited to low cell voltages and low energy densities. Because most of the electrolytes are based on AgI, the cells are restricted to potentials of less than 0.687 V. *Linford* et al. [4.20] recently reported solid state batteries based on silver sulphonium iodide electrolytes. They reported the cell

Cell VII $Ag/Ag_7(CH_2)_4SCH_3I_8/(CH_2)_4SCH_3I_3$

which exhibited a voltage of about 0.64 V. The electrolyte is a double salt that may be formed by the combination reaction

$$7AgI + (CH_2)_4SCH_3I \rightarrow Ag_7(CH_2)_4SCH_3I_8 . \tag{4.10}$$

The use of free I_2 as the cathode in Cell VIII was limited by the decomposition of the electrolyte in contact with I_2. *Linford* reported degradation of such cells by the rapid reaction

$$I_2 + Ag_7(CH_2)_4SCH_3I_8 \rightarrow 7AgI + (CH_2)_4SCH_3I_3 . \tag{4.11}$$

This results in a high internal resistance in the cell and a concurrent inability to deliver power. Other examples of this type of cell breakdown have been described for the degradation of Ag_3SI by I_2 [4.21]. *Linford* et al. reported that the solid electrolyte of Cell VII suffered a partial loss of conductivity following a two year period of storage at normal ambient temperature, but more promising results were reported for a silver iodide-bis-sulphonium diiodide solid electrolyte [4.20].

These observations are illustrative of certain important constraints on any solid electrolyte cell.

1) The high conductivity phase must be stable over the temperature range of application.

2) The electrolyte must not degrade by reaction with either electrode materials or environmental contaminants.

The review by *Liang* [4.1] provides an excellent discussion of the general aspects of solid state batteries including compatibility between electrodes and the electrolyte, the structure of electrodes and the interface between electrodes and electrolytes. The extensive work on Cell III addressed many of these problems, and will be discussed below.

Electrolyte Properties

Table 4.3 summarizes the significant properties of $RbAg_4I_5$. Electrical current is carried solely by the Ag^+ ion, electronic conductivity is negligible, the electrolyte is solid up to 232 °C and high ionic conductivity is exhibited over the practical ambient temperature range of nominally -55 °C to $+75$ °C.

The thermodynamic stability of $RbAg_4I_5$ is an important property that must be considered in any long term application. *Topol* and *Owens* [4.3] reported that $RbAg_4I_5$ is unstable at temperatures below 27 °C where it will disproportionate according to the reaction

$$2RbAg_4I_5 \rightarrow 7AgI + Rb_2AgI_3 . \tag{4.12}$$

Therefore, based upon thermodynamic considerations, the conducting phase would be nonexistent at low temperature. However, it was found that this reaction requires a catalyst such as H_2O vapor in order to initiate. By handling the materials in a dry atmosphere and hermetically sealing the batteries, the conducting phase has been maintained in devices for five years at temperatures below 27 °C [4.22, 23].

The synthesis of $RbAg_4I_5$ is straightforward. One method involves intimately mixing stoichiometric amounts of RbI and AgI, adding sufficient H_2O to form a thick fluid paste, and then removing the H_2O by air drying at slowly increasing temperatures, finally drying and combining the remaining reactant phases at a temperature of 210–229 °C, which is intermediate between the eutectic temperature and the incongruent melting point. By performing this reaction process over a two day period, single phase $RbAg_4I_5$ is formed. The material is then powdered and stored in a vacuum oven at $\sim 70\,^\circ C$ to prevent any environmental degradation.

A typical cell electrolyte is fabricated from the powdered $RbAg_4I_5$ by combining it with 5 to 10 w/o of a plastic, e.g., a polycarbonate resin. This is performed by dissolving the plastic in acetone, mixing it with the $RbAg_4I_5$, air drying the product and then powdering it again. In this manner an electrolyte powder is obtained which is readily formed into a dense pressed pellet for use in solid state batteries.

Electrodes

Silver must be used as the negative electrode (anode) with $RbAg_4I_5$, as previously discussed, because all of the electrical current is transported by Ag^+ ions. Silver foil is unsatisfactory because of the small area of the $Ag/RbAg_4I_5$ interface, which results in poor electrical contact and high polarization. Powdered silver, intermixed with $RbAg_4I_5$, provides increased capacity and lower resistance, but the efficiency (percentage utilization of the active component) of the electrode is still poor. An electrode formed by the *in situ* reduction of Ag_2O by carbon, intermixed with electrolyte, produced a high efficiency, low polarization electrode [4.24]. This type of silver electrode is readily pelletized and formed into a cell, and discharged efficiently over a wide temperature range.

The I_2 in the positive electrode (cathode) must be complexed to reduce its activity because elemental I_2 degrades the electrolyte via the reaction

$$I_2 + RbAg_4I_5 \rightarrow RbI_3 + 4AgI . \tag{4.13}$$

However, by complexing the I_2 with Me_4NI (Me = methyl), one forms a series of polyiodide salts [4.25, 26]

$$Me_4NI + I_2 \rightarrow Me_4NI_3 \tag{4.14}$$

$$Me_4NI + I_2 \rightarrow Me_4NI_5 \tag{4.15}$$

$$Me_4NI_5 + 2I_2 \rightarrow Me_4NI_9 . \tag{4.16}$$

These reactions are reversible and the salts are, therefore, sources of I_2 at reduced activity. Two cathodes that have been extensively investigated in batteries are Me_4NI_5 and Me_4NI_9. To provide both electronic and ionic conductivity within the cathode, carbon and $RbAg_4I_5$ are mixed with the polyiodide compound. A typical cathode mixture is fabricated by blending powdered carbon, $RbAg_4I_5$ and Me_4NI together with the stoichiometric amount of I_2 [4.27]. This mixture is slowly heated to $120\,^\circ C$ over a 24 h period, and then slowly cooled. The product is then ground to a coarse powder which can be readily pressed into a pellet for cell fabrication. Such cathodes have exhibited efficiencies in excess of 90%.

The important features in selecting a polyiodide cathode include consideration of its weight, stability, complexing ability and discharge reaction. A low weight is desired to maximize the energy density. The complexed I_2 should be readily available for discharge, preferably at a potential just under the decomposition voltage. Higher molecular weight tetraalkylammonium iodides form polyiodides with more tightly bound I_2, resulting in lower discharge voltages. Thus the use of tetrabutylammonium iodide (TBAI) resulted in cell voltages of 0.56 V, as reported by *de Rossi* et al. [4.28]. RbI was also considered as the complexing salt for cathodes [4.17]. RbI forms only the triodide RbI_3 whereas higher polyiodides could be formed by Me_4NI, thus favoring the latter as the donor component. The weight percent of available I_2 in RbI_3 is only 54%, whereas in Me_4NI_9 it is 83%.

However, the most significant property that leads to the selection of Me_4NI_5 and Me_4NI_9 as cathodes is the discharge reaction. If the discharge product is an ionically conducting solid electrolyte, then the electrode functions with low polarization and relatively thick high capacity electrodes may be discharged efficiently [4.17]. RbI forms a conductive phase when discharged at temperatures above 27 °C,

$$14Ag^+ + 7RbI_3 + 14e \rightarrow 3RbAg_4I_5 + 2Rb_2AgI_3 . \tag{4.17}$$

However, at lower temperatures, because $RbAg_4I_5$ is thermodynamically unstable, the net cathodic reaction is

$$4Ag^+ + 2RbI_3 + 4e \rightarrow Rb_2AgI_3 + 3AgI . \tag{4.18}$$

This results in increased cathode polarization and a loss of energy.

Me_4NI will combine with AgI to form the solid electrolyte $(Me_4N)_2Ag_{13}I_{15}$ [4.29]. Therefore, the net cathodic reactions for the two cathodes Me_4NI_9 and Me_4NI_5 are

$$16Ag^+ + 2Me_4NI_9 + 16e \rightarrow (Me_4N)_2Ag_{13}I_{15} + 3AgI \tag{4.19}$$

$$36Ag^+ + 9Me_4NI_5 + 36e \rightarrow 2(Me_4N)_2Ag_{13}I_{15} + 5Me_4NAg_2I_3 . \tag{4.20}$$

Assuming complete discharge, both (4.19) and (4.20) result in the formation of the ionically conductive $(Me_4N)_2Ag_{13}I_{15}$ as one of the components in the discharge product. This was confirmed by x-ray analysis of discharge products formed at low temperature (-40 °C). Stepwise discharges that could result in other intermediate products are apparent.

Cell Performance

Two modifications of Cell III will be discussed:

Cell VIII $Ag/RbAg_4I_5/Me_4NI_9$

Cell IX $Ag/RbAg_4I_5/Me_4NI_5$.

Table 4.4. Initial battery characteristics (25 °C) [4.31]

	$Ag/Me_4 NI_9$	$Ag/Me_4 NI_5$
Single cell voltage	0.662 V	0.650 V
Battery open circuit voltage	3.31 V	3.25 V
Internal resistance	30 Ω	35 Ω
Capacity (nominal)	40 mAh	40 mAh

The use of elemental I_2 in the cathode is not practical for the reasons previously discussed. The cell voltage of 0.687 exceeds the decomposition voltage of 0.66 and $RbAg_4I_5$ rapidly reacts with the I_2 as shown in (4.13). The open circuit cell voltages observed for Systems VIII and IX are given in Table 4.4; the single cell voltages do not exceed the decomposition voltage (0.66 V) and, therefore, the cells are stable.

The two modified Systems VIII and IX represent the culmination of several years of development effort. A number of papers report various aspects of these cell developments [4.2, 17, 22, 23, 30] and may be referred to for more detail. One configuration that was selected for advanced development was a 40 mAh capacity, anode limited cell, packaged as a series connected five-cell battery in a 0.5 inch diameter stainless steel can [4.31]. Each cell consisted of a solid three layer pellet 0.44 inch diameter by 0.12 inch height. The cells were connected in series and potted in the battery case with an epoxy compound. To prevent battery degradation by H_2O or air the battery was hermetically sealed.

The initial characteristics of these batteries are shown in Table 4.4. Because of the significant amount of internal resistance, these types of cells are best suited for low current (<1 mA) discharges where very stable operation over a wide temperature range is required. The batteries were discharged isothermally under constant loads of 21.5 kΩ and 64.9 kΩ. Figure 4.2 shows the discharge curves for the Ag/Me_4NI_9 batteries at temperatures of 71 °C, 23 °C, and -40 °C. It is significant that over this temperature range the discharge curves remained flat and the discharge life was relatively independent of temperature. In earlier versions of this battery it had been reported that at -55 °C less than 40% of the room temperature capacity was obtained [4.2]. The improvement in low temperature performance resulted from an optimization of the composition of the electrodes plus the removal of atmospheric impurities from the cell.

Both cell systems exhibited comparable discharge performances initially but the Me_4NI_5 cathodes polarized to a greater extent at -40 °C (as shown in Fig. 4.3) because the iodine is more tightly bound in the polyiodide. Another possible cause of greater polarization is that according to (4.19) and (4.20) Cell VIII will form a larger fraction of the conductive electrolyte $(Me_4N)_2 Ag_{13}I_{15}$ during discharge.

In summary the solid state batteries based upon Cell Systems VIII and IX discharge with a high efficiency over a wide temperature range, at low current densities. Problems associated with the solid state interfaces at the two electrodes

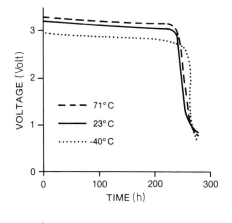

Fig. 4.2. 21.5 KΩ constant load discharge of Ag/Me$_4$NI$_9$ battery [4.31]

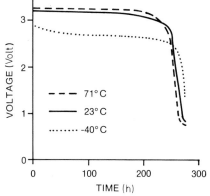

Fig. 4.3. 21.5 KΩ constant load discharge of Ag/Me$_4$NI$_5$ battery [4.31]

were avoided by the careful control of impurities and the use of high surface area electroactive materials intermixed with high surface area electronic conductors (carbon) and ionic conductors (RbAg$_4$I$_5$). The volumetric changes that take place in the electrodes because of the transport of Ag from the anode to the cathode were contained within the packaged cell by the case and encapsulating materials.

Storage Characteristics

An extremely desirable property of any primary battery is the capability to be stored without significant change in its energy delivering capacity. A battery is stored chemical energy wherein the chemical reactants combine electrochemically with the electron transfer from reductant to oxidant; this takes place external to the battery through a resistive load, as shown in Fig. 4.4. In this idealized cell the Ag and I$_2$ are combined electrochemically by the reaction

$$Ag + 1/2I_2 \rightarrow AgI \tag{4.21}$$

with all the electrons flowing through the external load R_L, and with the internal resistance of the cell being zero. In practice all primary batteries will exhibit some internal losses wherein one or both of the electroactive materials is consumed, as well as internal ohmic losses. This includes internal shorting, increased internal resistance and corrosion of the cell packaging materials.

The capability for a battery to be stored is referred to as its shelf life. In broad terms, the shelf life is the length of time that a battery may be stored and still deliver adequate power for a specific application. This is very subjective and will depend upon the application and also the storage conditions. One specific definition of shelf life is the length of time required for a battery on open circuit storage to lose ten percent of its initial capacity.

Self-discharge frequently involves the diffusion of active materials and is, therefore, dependent not only on temperature but also upon the physical state of the diffusing medium. One major advantage of a solid electrolyte battery over the more conventional liquid electrolyte batteries is its inherently lower self-discharge rate.

There are several different mechanisms for degradation of Cell VIII or IX.

1) The diffusion of H_2O and O_2 into the cell.

2) Recrystallization of the active silver in the anode to form larger crystallites with a lower surface area.

3) Electrolyte disproportionation to resistive phases at temperatures below 27 °C (4.12).

4) Iodine diffusion out of the cell and reaction with the encapsulant, current collectors or can materials.

5) Iodine diffusion through the electrolyte to the silver anode.

Atmospheric Degradation

The first mechanism is very serious for several reasons. The electrolyte disproportionation (4.12) is catalyzed by H_2O vapor and will readily take place below 27 °C. A second reaction that can occur at the $Ag/RbAg_4I_5$ interface is oxidation by O_2 and H_2O forming RbOH and AgI (4.5). Concurrent hydration of the RbOH can take place, with further degradation. A third method for anode degradation by H_2O and O_2 is a decrease in the electrochemical availability of the Ag resulting from surface contamination by H_2O and O_2. As previously stated, it is postulated that the adsorption of impurity species on a clean silver surface can inhibit surface self-diffusion of silver atoms, resulting in a lowered anode efficiency.

Silver Recrystallization

Silver recrystallization can take place during storage at elevated temperature. During the *in situ* reduction of Ag_2O to form the anode described above, the reactor was heated to several hundred degrees. If the reactor was not rapidly quenched when the Ag_2O reduction initiated, large particles of silver were

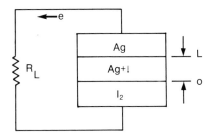

Fig. 4.4. Idealized solid state battery discharge for $Ag/AgI/I_2$ cell. Electrolyte thickness is L, and resistive load is R_L

formed and anode utilization was very low. However, by rapid cooling and subsequent storage at temperatures below 75 °C, recrystallization was not observed.

Electrolyte Disproportionation

Disproportionation according to (4.12) was not observed with $RbAg_4I_5$, provided the cell was stored hermetically. The 40 mAh batteries of Cell Types VIII and IX were tested after three years of storage at temperatures of 71 °C, 23 °C, and -15 °C [4.31]. The $RbAg_4I_5$ would be expected to disproportionate at -15 °C, causing an increase in the battery internal resistance. No such resistance degradation was observed and x-ray diffraction analysis of the anode and electrolyte layers confirmed that the $RbAg_4I_5$ was stable.

The electrolyte KAg_4I_5 may be used in place of $RbAg_4I_5$; it is isostructural with $RbAg_4I_5$ (see Sect. 3.1), exhibits the same value of ionic conductivity and is lower in cost. However, KAg_5I_5 is not as stable and disproportionates more readily than does $RbAg_4I_5$. KAg_5I_5 is thermodynamically unstable below 36 °C [4.3], and during low temperature storage of single-cell units at -15 °C, the electrolyte disproportionated significantly during the first two years [4.22].

Iodine Corrosion

The corrosion of cell and battery hardware is a concern in all batteries, and especially those in which storage at elevated temperatures for periods of several years is a requirement. Iodine resistant metals are available as are encapsulating materials. No significant loss in the capacity of the cathode due to this mechanism was observed during five years of storage at 71 °C or 23 °C. The cathode capacity losses observed during storage were due to intracell diffusion of iodine to the silver electrode.

Iodine-Silver Self Discharge

The only significant degradation mechanism that has been observed in hermetically sealed $Ag/RbAg_4I_5/Me_4NI_9$ batteries, during storage over the temperature range of -15 °C to 71 °C is the diffusion of iodine from the cathode to the silver anode. Early developmental model batteries lost nearly 50% of their capacity during two years of storage at 71 °C and 10% of their capacity at 23 °C [4.30]. The variation in capacity as a function of time is shown in Fig. 4.5, from which one concludes the shelf life is about 0.5 year at 71 °C and 2 years at 23 °C.

These batteries were cathode limited, with initial capacities of 50 mAh. During storage while the cathodes were losing capacity, the internal resistance of the batteries was also changing, increasing from an initial value of 30 Ω to values of several thousand ohms in one to two years, as shown in Fig. 4.6. Even within the first six months of 23 °C storage, the resistance increased to several hundred ohms, and caused unacceptable polarization losses in low temperature discharge. Both of these effects are readily explained by the same process, the diffusion of iodine to the anode.

The cell capacities were limited by the amount of iodine in the cathode. As I_2 diffuses out of the cathode, the cathode capacity (and, therefore, cell capacity) is reduced proportionately. In the cathode

$$Me_4NI_9 \rightarrow Me_4NI_5 + 2I_2 \tag{4.22}$$

and the I_2 released establishes a concentration gradient across the electrolyte layer resulting in a constant flux of I_2 from the cathode to the anode.

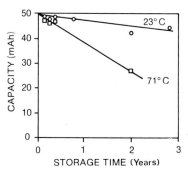

Fig. 4.5. Five-cell developmental model battery stack capacity to 2.7 V as a function of storage time [4.30]

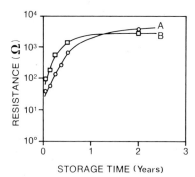

Fig. 4.6. Resistance of batteries at 23 °C, following isothermal storage: (*A*) stored at 23 °C, (*B*) stored at 71 °C [4.30]

At the anode surface the I_2 is reduced by the silver either chemically as shown in (4.21) or by local cell action,

$$I_2 + 2Ag^+ + 2e \rightarrow 2AgI \qquad (4.23)$$

with Ag from the internal anode structure being oxidized and ionically transported to the anode/electrolyte interface.

Three effects take place at the anode. First, the I_2 activity is effectively maintained at zero, and, therefore, there is a constant I_2 concentration gradient from the cathode to the anode to maintain a steady I_2 flux. Second, the anode loses capacity at the same rate as the cathode. Since the cathode was initially the limiting electrode, it still determines the cell capacity observed in Fig. 4.5. Third, the AgI is formed as a discrete layer between the anode layer and the electrolyte layer. The ionic conductivity of AgI at 23 °C has a range of values on the order of 10^{-4} to 10^{-6} $[\Omega\,cm]^{-1}$, whereas the composite battery electrolyte has a conductivity of 10^{-1} $[\Omega\,cm]^{-1}$. Therefore, as this layer of AgI nucleates and grows in a planar configuration, the internal resistance is increased by several orders of magnitude as shown in Fig. 4.6.

Oxley and *Owens* [4.17] determined the rate of diffusion of iodine through various electrolytes with the cell

Cell X   Carbon, $RbAg_4I_5$/$RbAg_4I_5$/I_2 (complexed).

              (I)   (II)   (III)

The carbon electrode (I) was biased at a potential negative to the iodine electrode (III), and as iodine diffused to electrode (I) it was reduced under limiting current conditions. Thus the steady state current flowing in this cell was a direct measure of the electrode capacity loss rate. The diffusion rate was found to be a function of electrolyte thickness, temperature and iodine activity. No iodine diffusion was detected through single crystal $RbAg_4I_5$. The diffusion is structure sensitive and is believed to occur along grain boundaries and microfissures in the polycrystalline, polyphase battery electrolyte.

There was good agreement between the observed cell capacity losses and those calculated from the iodine diffusion rates, supporting the conclusion that this is the only significant mechanism for self-discharge [4.17].

Oldham and *Owens* [4.23] have mathematically modeled this self-discharge mechanism. The cathode is assumed to be separated from the silver anode by a layer of electrolyte of constant thickness L, as shown in Fig. 4.4. The iodine concentration at the cathode/electrode interface ($x=0$) is constant at a value C_1. Assuming that Fick's second law is obeyed as the iodine diffuses to the anode surface ($x=L$),

$$D\frac{\partial^2 c}{\partial x^2} = \frac{\partial c}{\partial t} \qquad (4.24)$$

where D is the effective diffusion coefficient for I_2 in the electrolyte, c is the concentration at time t and distance x. At time $t=0$, the cell is first assembled, and there is no iodine in the electrolyte.

$$c=0, \quad x>0, \quad t=0. \tag{4.25}$$

The assumption of a constant iodine concentration at the cathode/electrolyte interface is expressed as

$$c=C_I, \quad x=0, \quad t>0. \tag{4.26}$$

At the anode/electrolyte interface the iodine is assumed to reduce instantaneously to AgI, and therefore,

$$c=0, \quad x=L, \quad t\geq0. \tag{4.27}$$

The solution to (4.24) for these boundary conditions can be expressed as

$$\frac{c}{C_I}=1-\frac{x}{L}-\frac{2}{\pi}\sum_{j=1,2\ldots}^{\infty}\frac{(-)^j}{j}\exp\left(\frac{-j^2\pi^2Dt}{L^2}\right)\sin\left[\frac{j\pi(x-L)}{L}\right]. \tag{4.28}$$

The flux J of iodine diffusing to the anode can be obtained by applying Fick's first law,

$$J=-D\left(\frac{\partial c}{\partial x}\right)_{x=L} \tag{4.29}$$

to (4.28). The result is

$$\frac{JL}{DC_I}=\frac{2L}{\pi^{1/2}D^{1/2}t}\sum_{j=1,3,5\ldots}^{\infty}\exp\left(\frac{-j^2L^2}{4Dt}\right). \tag{4.30}$$

The AgI was assumed to grow as a planar layer with density ϱ, conductivity κ and molecular weight M. The rate of increase of cell resistance R is expressed as

$$\frac{dR}{dt}=\frac{2MJ}{\varrho\kappa A} \tag{4.31}$$

where A is the area of the anode layer. The solution for the rate of resistance increase is

$$\frac{\varrho\kappa A}{2MLC_I}(R-R_0)=\frac{Dt}{L^2}+\frac{2}{\pi^2}\sum_{j=1,2\ldots}^{\infty}\frac{(-)^j}{j^2}\left[1-\exp\left(\frac{-j^2\pi^2Dt}{L^2}\right)\right] \tag{4.32}$$

where R_0 is the initial cell resistance.

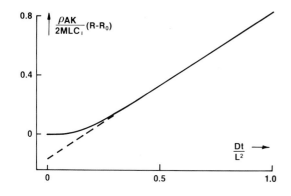

Fig. 4.7. Graph showing the build-up of cell resistance [4.23]

This result is shown in Fig. 4.7. This model predicts that there will be an initial period of constant resistance, followed by an increasing rate of resistance buildup, finally reaching a steady state wherein the resistance increases at a constant rate. The model assumes that the formation of the AgI layer does not affect the diffusion rate or the value of L. Experimental data are limited, but there appears to be qualitative agreement between the battery data in Fig. 4.6 and this model. The period of no increase in resistance is assumed to be very short, possibly ending before the battery is fabricated and, therefore, not reflected in Fig. 4.6. The curves in Fig. 4.6 show a period of several months duration during which resistance increased exponentially, followed by a reduced rate of increase to an approximately linear rate. The effective diffusion coefficients for iodine were reported [4.23] to be $3 \cdot 10^{-10}$ and $5 \cdot 10^{-10} \, cm^2 \, s^{-1}$ at temperatures of $23°\,C$ and $71°\,C$, respectively.

Because of the very significant degradation of the $Ag/RbAg_4I_5/Me_4NI_9$ batteries during storage, there was serious question about developing a viable solid state battery based upon any iodine complex cathode with a measurable vapor pressure of iodine. Three modifications to the cell design significantly improved the battery storage behavior.

1) Finely divided silver was dispersed into the electrolyte phase to act as an iodine getter. This was to prevent iodine from diffusing to the anode, until all of the getter was oxidized. Thus, this getter silver would delay the initiation of anode degradation [4.32].

2) The iodine activity was reduced by forming a stronger complex. Me_4NI_5 was used as a compromise. Stronger complexes could be formed, but they did not discharge as well because of concentration polarization resulting from reduced iodine mobility and ohmic polarization due to formation of more resistive discharge products.

3) Some excess RbI (in the form of Rb_2AgI_3) was incorporated into the electrolyte layers so that instead of AgI being formed by I_2 diffusion, the net self-discharge reaction becomes

$$7I_2 + 14Ag + 2Rb_2AgI_3 \rightarrow 4RbAg_4I_5 . \qquad (4.33)$$

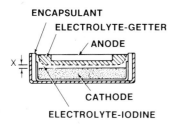

ENCAPSULANT

ELECTROLYTE-GETTER

ANODE

CATHODE

ELECTROLYTE-IODINE   Fig. 4.8. Cross section of cell assembly [4.23]

In this manner the resistive layer of AgI that previously formed at the $Ag/RbAg_4I_5$ interface is replaced by a layer of the conducting $RbAg_4I_5$ electrolyte [4.33].

Figure 4.8 is the cross section of a cell that incorporated these modifications [4.30]. Batteries based on both polyiodide cathodes were developed because of the initially higher rate capability of the Me_4NI_9 cathode relative to the Me_4NI_5 cathode. In these cells the cathode begins to lose capacity as soon as the cells are fabricated by reacting with the getter, whereas the anode capacity is constant until all the getter material has been oxidized. In this anode limited cell design, it is important to note that the iodine self-discharge reaction has not been eliminated, but, rather, has been modified so that there is no degradation in the impedance of the cell. The cell was initially anode limited with a 40 mAh anode. The cathode capacity of 50–54 mAh was sufficient to provide iodine for the normal electrochemical discharge of the anode silver and also for the degradation reaction with the getter silver. The resistive AgI formed during storage was dispersed throughout the electrolyte and, therefore, did not form an internal series resistance. Further, when the storage temperature exceeded 27° C, this AgI combined with the excess RbI to form $RbAg_4I_5$.

Oldham and *Owens* [4.23] mathematically modeled the cell for the design in which getter was incorporated into the electrolyte and the composition of the ionic component in the anode and electrolyte remained as stoichiometric $RbAg_4I_5$. Initially there is a uniform concentration C_A of the getter silver in the electrolyte zone from the cathode $(x=0)$ to the anode $(x=L)$. As diffusion takes place, a sharp boundary is formed in the electrolyte (at $x=X$) between the region near the cathode where the getter silver has been converted to silver iodide $(0<x<X)$ and the region near the anode containing the unreacted silver getter $(X<x<L)$. Figure 4.8 illustrates this boundary as a planar layer. The rate at which this plane $x=X$ moves toward the anode is proportional to the iodine flux J arriving at this interface. Therefore,

$$C_A\left(\frac{dX}{dt}\right) = 2J = -2D\left(\frac{\partial c}{\partial x}\right)_{x=X} \tag{4.34}$$

represents one boundary condition for Fick's second law (4.24) and the other boundary condition is (4.26). The initial conditions in this cell are represented by

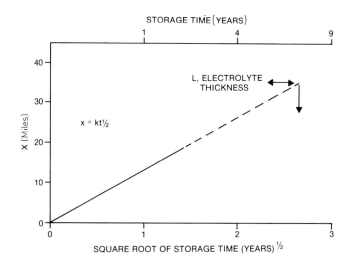

Fig. 4.9. Idealized $X-t^{1/2}$ function

(4.25). The approximate solution is

$$X \simeq \left(\frac{8C_1Dt}{C_A}\right)^{1/2}. \tag{4.35}$$

This model, thus, predicts that the rate at which the I_2/Ag boundary $(x=X)$ moves toward the anode $(x=L)$ will be proportional to the square root of time for batteries that are stored isothermally. The other terms of (4.35) are not known well enough to permit calculation of the $X-t$ function. However, by experimentally determining the slope of the $X-t^{1/2}$ function, one can predict how long it will take for the iodine diffusion distance τ to equal the electrolyte thickness L. This time interval τ is expressed as

$$\tau = L^2 \left(\frac{dX}{d\sqrt{t}}\right)^{-2}. \tag{4.36}$$

According to the present model, a plot of X versus $t^{1/2}$ should yield a straight line passing through the origin, as shown in Fig. 4.9. The solid line represents real data; extrapolation of this curve to $X=L$ yields the value of τ, the storage time during which the cell capacity should be constant.

Owens et al. [4.30] reported the values of X and t for Cell System VIII. The 40 mAh cells were isothermally stored at temperatures of $-15°$C, $23°$C, and $71°$C. Cells were destructively analyzed at time intervals of 0, 0.25, 0.5, and 1.0 year. The stored cells were electrically tested (voltage and resistance) and discharged to determine capacity. Then the discharged cells were sectioned and the iodine diffusion distance was measured with an optical microscope. This series of tests was continued for five years, and Owens [4.22] subsequently reported results for four types of cells. Table 4.5 identifies these four types which

Table 4.5. Experimental cell types [4.22]

Type	Anode	Electrolyte	Cathode
1	Ag	$RbAg_4I_5 + Ag$	$(CH_3)_4NI_9 + C + RbAg_4I_5$
2	Ag	$RbAg_4I_5 + Ag^a$	$(CH_3)_4NI_9 + C + RbAg_4I_5$
3	Ag	$KAg_4I_5 + Ag$	$(CH_3)_4NI_9 + C + KAg_4I_5$
4	Ag	$RbAg_4I_5 + Ag$	$(CH_3)_4NI_5 + C + RbAg_4I_5$

[a]   Modified process for getter addition.

include the two cathodes Me_4NI_9 and Me_4NI_5 and the KAg_4I_5 electrolyte as well as the $RbAg_4I_5$ electrolyte. The results for Cell Type 1 are shown in Fig. 4.10. Each data point is the average of at least two cells; generally the agreement was good to ± 0.001 inch. Because the cells were subjected to an initial temperature cycle as a part of the fabrication process, the first measurements on real cells were corrected for the hypothetical isothermal cell. In all cases the linearity of the $X - \sqrt{t}$ function was quite good for all four cell types and it was possible to obtain a reasonable prediction of τ from relatively short term storage tests.

The predicted values of τ for the four types of cells are given in Table 4.6. The predicted values for the $Ag/RbAg_4I_5/Me_4NI_9$ systems (Cell Types 1 and 2) are in reasonable agreement, indicating that ten years would elapse before any loss in cell capacity would be observed for cells stored at 23° C. The effective diffusion coefficient of I_2 through KAg_4I_5 is less than that through $RbAg_4I_5$, based upon measurements in cells similar to Cell X.

The results in Table 4.6 for the $Ag/KAg_4I_5/Me_4NI_9$ cell (Type 3) indicate that 40 years are required for the diffusing iodine to reach the anode at 23°C, a factor of 4 improvement in storage life relative to the comparable $RbAg_4I_5$ systems. This indicates that the effective iodine diffusion coefficient in $RbAg_4I_5$ is 16 times greater than in KAg_4I_5.

The use of Me_4NI_5 in place of Me_4NI_9 increases the value of τ by a factor of 2, as shown in Table 4.6. This is in agreement with the anticipated reduced I_2 flux

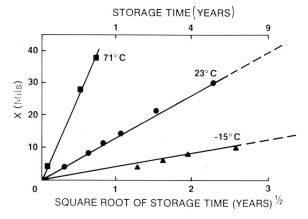

Fig. 4.10. Graph of X versus $t^{1/2}$ for $Ag/RbAg_4I_5/Me_4NI_9$ cells [4.23]

Table 4.6. Predicted values of τ in years [4.22]

Cell type	τ −15 °C	τ 23 °C	τ 23 °C
1	100	9	0.6
2	160	11	0.8
3	400	40	2.4
4	200	20	1.7

Table 4.7. Solid electrolyte cell capacity following storage [4.22]

Cell type	Storage T [°C]	Cell capacity [mAh]			
		Storage time			
		zero	1 year	2 years	5 years
1	71	39	37	29	15
	23	39	39	39	39
	−15	39	39	39	38
2	71	39	37	32	—
	23	39	40	39	39
	−15	39	40	—	38
3	71	40	39	38	33
	23	40	41	—	—
	−15	40	40	36	30–39
4	71	40	39	40	30
	23	40	40	41	40
	−15	40	41	—	—

due to the lower I_2 vapor pressure in the Me_4NI_5 cathode. The factor of 2 implies a ratio of 4 between the values of C_I in the Me_4NI_9 cathode and the Me_4NI_5 cathode. These predictions of cell capacity are in reasonably good agreement with other measurements of cell degradation. Table 4.7 reports the cell capacities initially and after storage [4.22].

The capacities of Type 1 cells, following storage of up to five years, are plotted in Fig. 4.11. Cells stored at 71° C exhibit no capacity loss during the first 0.5 year, in good agreement with the τ value of Table 4.6. Cells stored at 23° C or −15° C have no significant loss after five years [4.22].

The cell resistances for Type 1 are shown in Fig. 4.12. At 71° C, resistance increase initiated after a storage interval of 0.5 year. At −15° C there was no change in five years. Cells at 23° C exhibited some resistance increase after about four years. This is in disagreement with the cell capacity and sectioning measurements, and may be caused by the diffusion of H_2O and O_2 into the cell. These cells were not truly hermetically sealed as they were potted with epoxy.

Warburton and *Owens* [4.31] are investigating multicell batteries based on Systems VIII and IX. These batteries are hermetically sealed, utilize the silver getter and also have excess RbI in the electrolyte phase. Following three years of

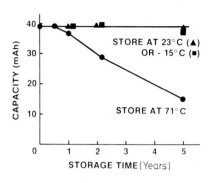

Fig. 4.11. Capacity of $Ag/RgAg_4I_5/Me_4NI_9$ cells [4.23]

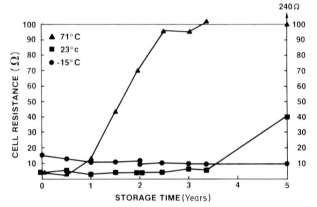

Fig. 4.12. Resistance of $Ag/RbAg_4I_5/Me_4NI_9$ cells stored at three temperatures [4.23]

storage the batteries are performing in agreement with predictions based upon the above described models for single cell storage behavior.

The investigations are summarized as follows:

1) The major cause of storage losses in hermetically sealed solid state cells of the type $Ag/RbAg_4I_5/Me_4N(I_9$ or $I_5)$ is iodine diffusing through the electrolyte to the anode.

2) The battery storage behavior can be mathematically modeled to give a theoretical basis for the prediction of shelf life.

3) Methods proposed for the control of self-discharge have been confirmed by five year tests.

4) KAg_4I_5 degrades at low temperatures and is not acceptable for long term storage applications.

5) The batteries based on $Ag/RbAg_4I_5/Me_4NI_5$ cells are predicted to be storable for 20 years at $23°C$ with no degradation in capacity or resistance.

4.3.2 Lithium-Metal Salt Solid State Batteries

Liang et al. have investigated solid state batteries based upon lithium anodes and various nonvolatile, metal salts as the cathode [4.1, 9, 10]. The use of elemental iodine or polyiodide compounds as a cathode has the disadvantages associated with the volatility of the iodine, as described in Section 4.3.1; the self-discharge and materials corrosion problems were caused primarily by the diffusion of the reactive iodine. These specific problems are avoided by the use of solid, nonvolatile salts as cathodes in high voltage cells [4.1].

The cell Li/LiI/AgI has an open circuit voltage of 2.1 V corresponding to the reaction [4.34]

$$Li + AgI \rightarrow LiI + Ag. \tag{4.37}$$

These cells were restricted to very low drain rates ($\sim 10\,\mu A/cm^2$) because of ohmic polarization losses. In addition the system was unstable because of interdiffusion of Li^+ and Ag^+ ions, resulting in self-discharge losses [4.35].

The use of the more conductive LiI (Al_2O_3) electrolyte permitted development of batteries based on Cell IV. *Liang* et al. reported a practical energy density of $0.5\,Wh/cm^3$ at low current densities [4.10].

Such batteries are stable and exhibit no capacity loss during storage at elevated temperatures. This system is possibly the most stable active primary battery available, provided it is hermetically sealed. Because of the reactivity of the Li with H_2O, N_2, and O_2 and the LiI with H_2O, the cells are fabricated in He dry boxes with the concentration of H_2O and O_2 being maintained at below 15 ppm [4.10].

One problem of a completely solid state battery is the volume change associated with cell discharge and the concurrent transport of matter. If the electrodes and electrolyte are relatively hard, brittle materials, the stresses developed during discharge will crack the cell, possibly causing internal shorting. High capacity cells of Type VIII (cathode thickness greater than 1 cm) cracked during discharge tests. *Doty* et al. [4.36] reported cracking in a limited number of Li/PbI_2 cells during isothermal discharge tests at 37° C. *Liang* has reported that redesign of the Type V cells has effectively solved this problem [4.37].

The batteries based on Cell IV are useful for low rate applications at normal or elevated temperatures. Improvements in capacity were recently reported; the use of As_2S_3 as the cathode depolarizer provides an energy density of 0.7 to 0.9 Wh/cm^3 [4.10].

4.3.3 Lithium-Halogen Solid Electrolyte Batteries

The batteries described in Sections 4.3.1 and 4.3.2 are based upon a pellet type of cell construction wherein anode and cathode disks are separated by a dense electrolyte disk of 0.02 to 0.10 cm thickness. The electrolyte is formed during fabrication as an integral layer separating the electrodes. Lithium-halogen batteries have been developed wherein the lithium halide electrolyte is formed *in situ* by the direct combination of the active electrode components. Cells of this type have been reported for Li/I_2 [4.38–41] and Li/Br_2 [4.42].

Pure iodine does not have adequate electronic conductivity to sustain discharge in a cell $Li/LiI/I_2$. However, by combining the I_2 with poly-2-

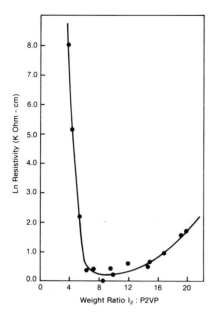

Fig. 4.13. Resistance of I_2-poly-2-vinylpyridine complexes as a function of composition at 37 °C [4.41]

vinylpyridine, one obtains a charge-transfer complex with an electrical conductivity of 10^{-3} $[\Omega\,cm]^{-1}$, as shown in Fig. 4.13 [4.41].

The Li/I_2 cell has been under development for several years and is presently finding successful application as the power source for cardiac pacemakers. The uniqueness of this battery is closely related to the composition of the cathode; an iodine poly-2-vinylpyridine charge-transfer complex (CTC) serves as the cathode. It is formed by combining the iodine and polymer to obtain a semisolid polymeric product containing about 92–94 w/o I_2. Significant properties of this CTC are its physical state, the ready availability of the I_2 and its electronic conductivity [4.41]. When this CTC contacts the Li anode, it immediately forms a thin film of LiI at the interface of the electrodes. The cell will promptly exhibit a cell voltage of 2.80, in good agreement with the value 2.796 V calculated from the Gibbs free energy of formation for LiI. This indicates that the cell reaction may be written simply as

$$2Li + I_2 \rightarrow 2LiI. \tag{4.38}$$

Although the I_2 is complexed (the nature of this complex has not been identified), it is probable that the iodine is at unit activity, as evidenced by the cell voltage. Discharge of the Li/I_2 – CTC cell according to (4.38) results in the formation of resistive LiI, and the battery impedance increases. The cell is, therefore, restricted to low current densities and moderate temperatures. It is ideally suited for the pacemaker application at 37° C and current densities of 1–10 $\mu A/cm^2$; under these conditions energy densities of as high as 0.8 Wh/cm^3 have been projected (Wilson Greatbatch, Ltd. Model 755 battery).

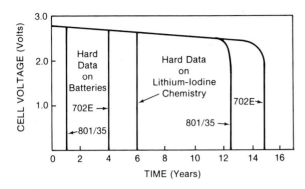

Fig. 4.14. Projected discharge curves for Li/I_2-CTC batteries at $37°C$ [4.41]

This Li/I_2 cell would be expected to undergo self-discharge by the diffusion of I_2 through the electrolyte to the anode, as was described for the Ag/I_2 cell (Sect. 4.3.1). *Kraus* and *Schneider* [4.41] estimate that a battery under discharge at $37°C$ will lose no more than 10% of capacity over a ten year period. Figure 4.14 shows discharge curves for Li/I_2 batteries in which operating times of well over ten years are projected for $37°C$ operation.

Greatbatch et al. [4.42] recently described a Li/Br_2-CTC cell that is similar to the Li/I_2-CTC cell. The solid electrolyte is LiBr and the cell exhibits a voltage of 3.50 V, again in good agreement with the thermodynamic value. The more energetic bromine cell is projected to deliver $1.25 Wh/cm^3$ at low rates. Few data are available for this system at present; however, based upon accelerated test results [4.43], projected curves for the W. Greatbatch, Ltd., iodine and bromine cells are compared in Fig. 4.15. The solid curves are based on accelerated discharge at high currents followed by equilibration at the indicated loads. The Br_2 cells were, therefore, discharged only out to 1.2 Ah, and the dashed curve represents a projection of anticipated performance assuming electrode and energy efficiencies comparable to the I_2 cell. The curves are for two cells of the same volume and correspond to energy densities of 0.8 and $1.25 Wh/cm^3$, respectively, for Li/I_2 and Li/Br_2.

The lithium/halogen cells represent a significant new type of power source. The Li/I_2-CTC batteries are extremely useful in powering cardiac pacemakers efficiently and with a high reliability. The number of other applications appears to be limited by the high impedance of the electrolytes, but these pacemaker cells demonstrate that resistive electrolytes (see Table 4.1) can be incorporated into useful electrochemical devices.

Owens and *Hanson* [4.11] developed cells of the type Li/LiI, Al_2O_3, H_2O/Me_4NI_5 with cell voltages of 2.75 V and energy densities (projected) of $0.40 Wh/cm^3$ at ten year discharge rates. They obtained cathode efficiencies of 95–100% in accelerated tests. These cells had a finite electrolyte layer fabricated as a discrete element and were subject to internal lithium shorts to the cathode. A unique feature of the Li/I_2-CTC cell is that if Li is shorted through the electrolyte to the cathode, it simply reacts to form a film of LiI and effectively eliminates the short. This fortuitous result removes one important failure

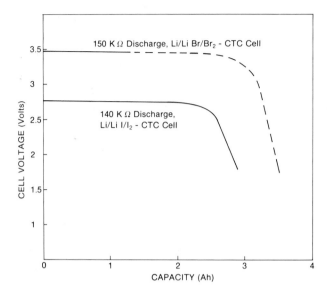

Fig. 4.15. Discharge curves for lithium/halogen cells at 37 °C [4.43]

mechanism from the system, thus increasing its reliability for long-term applications. Increased usage of the Li/halogen-CTC cells is anticipated in the future for low power applications.

4.4 Electrochemical Capacitors

The capacitance in an electrochemical capacitor derives from the electrical double layer that exists at an electrode/electrolyte interface. The values of capacitance that can be achieved are thus intrinsically higher than those in which a dielectric layer is employed and this constitutes their main advantage over dielectric types, i.e., a much higher capacitance per unit volume. A penalty incurred by elimination of a separate dielectric layer is that the maximum voltage per cell is considerably reduced, i.e., to below the electrolyte decomposition voltage (in practice <0.5 V for $RbAg_4I_5$ because a carbon electrode ceases to behave capacitively at greater potentials). This feature clearly has a strong influence on circuit design, although it is not a major shortcoming because the waveform can be readily amplified.

The basis for electrochemical capacitors utilizing $RbAg_4I_5$ is the ideally polarizable carbon electrode. The double layer capacity at a carbon/$RbAg_4I_5$ interface falls in the range $20\text{–}40\,\mu F/cm^2$ (real interface area). A dramatic multiplication in capacitance results when this electrode is fabricated as a structure with a high roughness factor, e.g., in a carbon/$RbAg_4I_5$ composite. Capacitances of up to 7 farads per gram have been achieved using high surface area carbons, corresponding to an effective interface area of $\sim 10^5 cm^2/g$. Devices of varying capacity can thus be fabricated simply by varying the weight of this electrode.

Both polar and nonpolar configurations of these capacitors have been developed. In the polar version represented by the cell,

Cell XI $Ag/RbAg_4I_5/C$

only the carbon electrode is capacitive. Silver on the other hand behaves as an ideally nonpolarizable electrode, polarization effects at this electrode being minimized, as discussed earlier, by use of a finely divided mixture of carbon, silver and the solid electrolyte. A nonpolar version would simply have two carbon electrodes

Cell XII $C/RbAg_4I_5/C$.

The inherent advantage of the nonpolar version results from absence of the silver electrode which can limit the useful life of the nonpolar version as a result of dendrite growth and the development of poor silver utilization, i.e., the silver electrode can cease to become ideally nonpolarizable. However, a nonpolar device has a perhaps more severe shortcoming resulting from the fact that the "potential of zero charge" or open circuit voltage of the carbon electrode is determined by the impurity level. Thus, whereas ideally the carbon electrode potential would be poised at a potential approximately equidistant between the silver and iodine potentials, in fact the potential in real cells is found to vary anywhere between 50 and 300 mV.

4.4.1 Capacitor Characteristics

In its polar form an electrochemical capacitor based upon $RbAg_4I_5$ is represented by Cell XI. During charge, Ag^+ ions migrate from right to left and discharge to form a layer of silver metal at the bulk silver electrode/electrolyte interface. The voltage change at this electrode is negligible compared with the voltage change at the carbon electrode, and it can thus be considered to have infinite capacity (cf. the negative electrode in a polar tantalum capacitor). The voltage change of the carbon electrode is nearly linear (at constant applied current) between 0 and 0.45 V, and the integral capacity (C_{int}) is thus given simply by the equation, $C_{int} = i dt/dE$. During discharge, exactly the reverse takes place, silver metal is oxidized to Ag^+ ions and the double layer at the carbon electrode is discharged. The separate charge/discharge characteristics for each electrode are plotted together in Fig. 4.16. Above about 0.5 V a pronounced curvature is observed (not shown in the figure) as a result of the onset of electrolyte decomposition. This, therefore, sets an arbitrary limit for operation of such devices. Devices of different capacity can be fabricated simply by varying the weight of the carbon electrode, which as discussed previously consists of a carbon/$RbAg_4I_5$ mixture.

Electrically, as shown in Fig. 4.17, an electrochemical capacitor can be represented as a two terminal network consisting of a large capacity in series

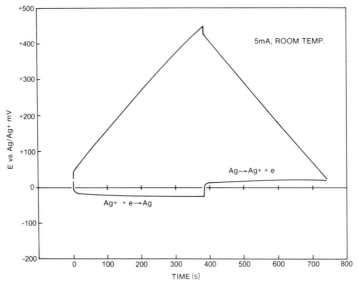

Fig. 4.16. Half-cell charge/discharge of 5 farad polar electrochemical capacitor [4.44]

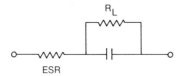

Fig. 4.17. Equivalent circuit for the solid electrolyte capacitor

with a low value resistor (ESR) and shunted by a very high value resistor (R_L). Values of R_L are greater than 10^{10} Ω. The equivalent series resistance (ESR) is higher than in conventional capacitors as a consequence of the bulk resistivity of the solid electrolyte as well as the "contact" resistances of the electrode materials. Typical values of ESR for one inch diameter, cylindrically packaged single cells are ~ 1 Ω at room temperature (the ESR decreases with increase in temperature). This, of course, limits the ability of the device to give up energy to an external circuit.

Table 4.8 presents comparisons of solid state electrochemical capacitors versus some more familiar capacitors [4.44]. Although such comparisons are useful from the standpoint of DC applications, they neglect the poor AC characteristics of electrochemical capacitors. It should thus be noted that the effective capacitance of these devices decreases with increasing frequency and the effective capacitance presented to a 20 Hz signal is two orders of magnitude less than that at DC [4.44].

Leakage currents are extremely small as a consequence of the low electronic conductivity of Ag^+ ion conductivity electrolytes, so long as adequate precautions are taken to exclude impurities, e.g., O_2 and H_2O which would faradaically discharge (or charge) the electrical double layer.

Table 4.8. Capacitor properties [4.44]

Capacitor type	Rated voltage [V]	CV [C/in^3]	$1/2\,CV^2$ [J/in^3]
Electrochemical (Ag$^+$ ion)	0.5	80	20
Tantalum wet slug	6	0.063	0.19
Aluminum	5	0.044	0.11
Ceramic	1000	0.0016	0.08

The device has a positive coefficient of capacitance in the order of 0.2% $(^\circ C)^{-1}$ over the temperature range -55 to $\sim 20\,^\circ C$. At higher temperatures, a lower coefficient is observed. The origin of these marked variations with temperature is probably related to mechanical effects associated with expansion and contraction of the carbon/RbAg$_4$I$_5$ interface, since true double layer capacity values are intrinsically less temperature sensitive.

4.4.2 Cycleability

The main criterion governing usefulness as an electronic component is the ability to undergo repeated cycling without change in electrical characteristics. Two failure mechanisms have been encountered with the polar configuration, namely silver dendrite growth and anode failure. Neither of these mechanisms is, of course, possible with a properly operated nonpolar device and this is clearly then the major advantage of this configuration.

There is, however, a region of low current density and low capacity density (CV/in^3) where apparently indefinite cycle life can be achieved for, e.g., a 0.5 farad device cycled between 0 and 0.5 V at 1 mA/cm^2. An important design consideration in avoiding dendrite growth is that the silver electrode has the larger diameter of the two electrodes so that the edge effect is avoided.

4.4.3 Applications

The high energy storage capability of the solid state electrochemical capacitor coupled with very low leakage current makes it an ideal device for use with operational amplifiers. Because of the low voltage at which charge is stored, the device can be charged from a voltage source through a resistor at what approximates very closely a constant current charge condition. Thus, 5 V and above power supply operation is totally acceptable. Because of the repeatable sawtooth waveform thus obtained over a very wide range of time delays, the main applications for which the devices are suited are in the areas of timing, integration and memory. The basic circuit approach on which such applications are based consists of a simple multivibrator or sawtooth generator using a single operational amplifier [4.45]. In timer applications, the actual times which are available are several orders greater than can be achieved with conventional capacitors, i.e., hours or days compared to seconds. Its integrating capability derives directly from the highly repeatable constant current charge/discharge characteristic; thus many timing points may be selected from one basic timing unit which enables the buildup of as many time sequences as are required for any multitime process control or monitoring system. Clearly, integration of both a linear and discontinuous nature is practical. The device can also be used as a digital memory. Here a fully charged device could represent a logic one and a fully discharged device a logic zero. The device can be discharged to 0 V or shorted out without affecting the capacitive carbon electrode because of the nucleation overpotential for deposition of silver onto carbon. Possible uses as a memory element are in industrial control systems, weapons systems and any other areas where an inexpensive, programmable, simple memory is required.

4.5 Solid Electrolyte Rechargeable Batteries

In many solid state applications, failure due to volatility of semiconductor systems during transient or long-term power failures may be in some cases inconvenient, in others hazardous or expensive. There is thus a need for a solid state, wide temperature energy store to be compatible with rugged, high reliability semiconductor electronic systems. As a source of secondary power, conventional secondary batteries have generally proven to be incompatible with semiconductor electronics. A useful device for these applications is a capacitive type energy store with all the environmental and shelf-life characteristics of a good quality capacitor, but at the same time an energy storage capability approaching that of a secondary battery. However, because of the low power requirements of modern semiconductor circuitry (e.g., complementary MOS), the specific energy storage capability required is considerably lower than that afforded by conventional secondary batteries.

In addition to standby power there are other potential applications, e.g., where it is necessary to minimize power supply requirements such as in a system employing a low current power source which must have an occasional low duty cycle, higher current output. A typical application is a telemetry system with occasional operating periods. Power could be obtained from solar cells or similar low power devices and stored in a rechargeable battery for use during an occasional data transmission.

The capacitor described previously can conceptually be employed in such applications but applicability is limited by its low energy storage capability and it is desirable that this be increased considerably. It should also be added that in applications of this type it is necessary that several devices be connected in series to provide a higher operating voltage.

As outlined previously, rechargeability of solid state cells based on Ag^+ ion conductors is limited by the buildup of high resistance of the positive electrode at potentials above 0.66 V as well as dendrite growth and poor utilization of the silver electrode. It has been shown [4.44] that the polar capacitors (described in the previous section) can be overcharged into the "pseudocapacity" region $0.5\,V < E < 0.66\,V$, e.g., to $\sim 0.625\,V$ and (depending on capacity) give several thousand cycles prior to failure from silver dendrite growth or the onset of poor silver utilization. As an example, a device with a rated capacity of 5 farads at 0.5 V, (0.7 mAh) when charged to 0.625 V gives ~ 3.5 mAh. This corresponds to approximately a tenth of a monolayer of iodine adsorbed at the carbon/electrolyte interface. At this level, the iodine is firmly bound and a device charged to 0.625 V retains its charge when the current is interrupted. So long as care is taken during fabrication that close tolerance is maintained of the weight of the carbon/$RbAg_4I_5$ mixture, which governs the amount of charge which the devices will accept, no problems are encountered in the electrical symmetry of multicell series stacks.

These devices fail as a result of gradual deterioration of the silver electrode which is governed partly by the thickness of the silver deposit (mAh/cm^2) and the charge and discharge current densities. This can take up to several thousand cycles for devices of relatively low capacity, e.g., a 5 farad device charged to 0.625 V (3.5 mAh).

Higher capacity devices or those charged to a higher voltage will have lower cycle life but this may be tolerable, e.g., in standby power applications requiring only a few cycles.

4.6 Solid Electrolyte Coulometers (Timers)

Coulometry based on the electrodeposition of a metal and subsequent anodic stripping from an inert substrate has long been a viable electroanalytical tool. The basic requirement is that the coulombic efficiency be unity or close to unity. Liquid electrolyte systems, usually based on silver, have a long record of satisfactory use in such applications under normal ambient conditions [4.46, 47]. However, at extremes of temperature or where the application demands unusual ruggedness, reliability and accuracy, it is desirable not to rely on a liquid electrolyte system. Typical instances are when the device must be incorporated as an integral component in microelectronic circuitry or be subjected to unusual stress as in certain military uses, e.g., timing devices for ordnance fusing. Numerous commercial uses have been foreseen for coulometers of both the liquid and solid electrolyte types. Examples of such uses include engine and equipment maintenance, warranty monitoring and virtually any other area where timing or integration of an electrical signal is required.

The electrical response or voltage/time characteristic of the timers discussed here differs from that of the capacitors discussed previously in that during "time out" when the silver (or other metal) is anodically stripped and replated on the counter electrode, the voltage remains low until all the metal is consumed. Then the voltage rises abruptly, at a rate determined in the limit by the double layer capacity of the substrate material, to a value determined by the potential of the next possible electrochemical reaction. This change is sufficient to trigger, for example, an SCR and cause the appropriate electronic signal to be generated.

A variety of different solid state timer cell materials and configurations has been developed into practical devices. One such device which has achieved commercial success has been developed by Sanyo in Japan and is sold under the trade name of "Couliode". This is presumably based on the cell

$$Ag/Ag_2S, \ Ag_2HgI_4/Au.$$

A device based on $RbAg_4I_5$ may be represented by the same cell scheme as the polar capacitor discussed previously, i.e., Cell XI. The only difference is in the construction and mode of operation of the carbon electrode. (Other inert materials can be used instead of carbon but the latter is preferred from the standpoints of ease of construction and cost.) This device can be set as a timer by plating a layer of silver metal onto the carbon surface. Operating now in reverse, the silver is stripped until it is entirely removed when the voltage will rise and timeout occurs. There are two requirements for the successful performance of such devices, first that all of the plated silver be available for subsequent "stripping" and second that faradaic impurities be excluded from the system. The latter is, of course, particularly important for small coulombic charges or when it is required that the device have an extended stand-life in the pre-set condition. To avoid "loss" of silver, which is apparent in poor coulombic efficiency, it is essential that, when the device is fabricated, good electrode/electrolyte contact is established. Where regions of poor contact exist, the silver can become detached from ionic or electronic contact at the interface and will not be available for oxidation. One way of ensuring adequate contact has been to form and package the devices under a positive pressure.

Kennedy et al. [4.48, 49] have reported the development of thin film timing devices of this type. Cells of the type $Ag/RbAg_4I_5/Au$ were prepared. The $RbAg_4I_5$ was formed by condensation on the substrate after evaporation from an 80 m/o AgI, 20 m/o RbI mixture in an evaporation boat; electrolyte thickness in these cells varied from 80000 Å to 200000 Å. Conductivities found were 0.25 $[\Omega \text{ cm}]^{-1}$, in close agreement with values obtained with polycrystalline materials. The accuracy of the $RbAg_4I_5$ cells was greater than 99% at current densities of 10 µA and for charge times between 20 and 250 s. At current densities of only 1 µA, however, deviations in accuracy of nearly 30% occurred. Cells were cycled on a continuous basis with 200 µC of charge for over 1000 cycles with no apparent cell degradation. Cell life was in fact limited by the loss of contact between the gold and $RbAg_4I_5$ electrolyte. This effect was presumably enhanced by the presence of atmospheric contaminants.

Note Added in Proof

Recently *Sanyo* [4.50] have introduced a low capacity solid electrolyte rechargeable battery represented by the concentration cell,

$$Ag/Ag_6I_4WO_4/Ag_2Se\text{-}Ag_3PO_4.$$

The activity of silver in the positive electrode is changed by the passage of current up to a maximum cell voltage of 0.280 V. Between 0 and 0.120 V the voltage change with Coulombs is surprisingly linear. This feature has been adapted by Sanyo for use in timing, integrating and memory applications. Devices are sold under the trade name of "Memoriode". They have also been constructed with a third electrode as a reference for potential sensing thus avoiding over-potential effects associated with the silver electrode. In terms of electrical characteristics and potential applications these devices have much in common with the electrochemical capacitors described in Section 4.4. They would appear, however, on the basis of published information, to offer somewhat better temperature characteristics although the voltage linearity is not so good.

References

4.1   C.C.Liang: Solid State Batteries. In *Applied Solid State Science*, Vol. **4** (1974) p. 95
4.2   B.B.Owens: Solid Electrolyte Batteries. In *Adv. in Electrochem. and Electrochem. Eng.*, Vol. **8**, ed. by C.W.Tobias (J. Wiley and Sons, Inc., New York 1971) Chapt. 1, pp. 1–62
4.3   L.E.Topol, B.B.Owens: J. Phys. Chem. **72**, 2106 (1968)
4.4   A.F.Sammells, J.Z.Gougoutas, B.B.Owens: J. Electrochem. Soc. **122**, 1291 (1975)
4.5   T.G.Stoebe, P.L.Pratt: Proc. Brit. Ceram. Soc. **9**, 181 (1967)
4.6   D.C.Ginnings, T.E.Phipps: J. Am. Chem. Soc. **52**, 1340 (1930)
4.7   C.R.Schlaikjer, C.C.Liang: J. Electrochem. Soc. **118**, 1447 (1971)
4.8   C.C.Liang: J. Electrochem. Soc. **120**, 1289 (1973)
4.9   C.C.Liang, L.H.Barnette: J. Electrochem. Soc. **123**, 453 (1976)
4.10 C.C.Liang, A.V.Joshi, L.H.Barnette: Proc. 27th Power Sources Symposium, Atlantic City, NJ 1976, p. 141
4.11 B.B.Owens, H.J.Hanson: U.S.Patent No. 4007/22 (1977)
4.12 Y.Sauka: J. Gen. Chem. **19**, 1453 (1949)
4.13 C.Tubandt: Z. Anorg. Chem. **115**, 105 (1921)
4.14 J.H.Kennedy, R.Miles, J.Hunter: J. Electrochem. Soc. **120**, 1441 (1973)
4.15 J.Schoonman, G.J.Dirksen, R.W.Bonne: Solid State Commun. **19**, 783 (1976)
4.16 J.Schoonman: J. Electrochem. Soc. **123**, 1772 (1976)
4.17 J.E.Oxley, B.B.Owens: Solid State Batteries. In *Power Sources 3*, ed. by D.H.Collins (Oriel Press, Newcastle Upon Tyne, England 1971) p. 535
4.18 D.O.Raleigh: J. Phys. Chem. **71**, 1785 (1967)
4.19 J.E.Oxley: Unpublished work (1967)
4.20 R.G.Linford, J.M.Pollock, C.F.Randell: Proc. 10th International Power Sources Symposium, Brighton, England, 1976
4.21 B.B.Owens, G.R.Argue, I.J.Groce, L.D.Hermo: J. Electrochem. Soc. **116**, 312 (1969)
4.22 B.B.Owens: Five Year Storage Tests of Solid State Ag/I$_2$ Cells, 148th Nat. Meeting Electrochem. Soc., 1975, Abstr. No. 43
4.23 K.B.Oldham, B.B.Owens: Electrochim. Acta **22**, 677 (1977)
4.24 I.J.Groce: U.S. Patent No. 3503810 (1970)
4.25 H.W.Foote, M.Fleischer: J. Phys. Chem. **57**, 122 (1953)
4.26 L.E.Topol: Inorg. Chem. **10**, 736 (1971)
4.27 B.B.Owens: J. Electrochem. Soc. **117**, 1536 (1970)

4.28 M.DeRossi, G.Pistoia, B.Scrosati: J. Electrochem. Soc. **116**, 1642 (1969)

4.29 B.B.Owens: J. Electrochem. Soc. **117**, 1536 (1970)

4.30 B.B.Owens, J.S.Sprouse, D.L.Warburton: Proc. 25th Power Sources Symposium, Atlantic City, NJ 1972, p. 8

4.31 D.L.Warburton, B.B.Owens: to be published

4.32 B.B.Owens, J.R.Humphrey: U.S. Patent No. 3661647 (1972)

4.33 B.B.Owens, J.E.Oxley: U.S. Patent No. 3663299 (1972)

4.34 C.C.Liang, P.Bro: J. Electrochem. Soc. **116**, 1323 (1969)

4.35 C.C.Liang: J. Electrochem. Soc. **118**, 894 (1971)

4.36 R.L.Doty, K.Fester, T.Kuder, W.Tracinski: "Lithium Batteries: Are They All the Same?", 5th International Symposium on Cardiac Pacing, March 14–18, 1976, Tokyo, Japan, to be published

4.37 C.C.Liang: private communication (1976)

4.38 A.A.Schneider, J.R.Moser, T.H.E.Webb, J.E.Desmond: Proc. 24th Power Sources Symposium, Atlantic City, NJ, 1970, p. 27

4.39 W.Greatbatch, J.H.Lee, W.Mathias, M.Eldridge, J.R.Moser, A.A.Schneider: IEEE Transactions on Bio-Medical Eng. BME-**18**, 317 (1971)

4.40 A.A.Schneider, W.Greatbatch, R.Mead: Performance Characteristics of a Long Life Pacemaker Cell. In *Power Sources 5*, ed. by D.H.Collins (Academic Press, London 1975) p. 651

4.41 F.E.Kraus, A.A.Schneider: Proc. 27th Power Sources Symposium, Atlantic City, NJ, 1976, p. 144

4.42 W.Greatbatch, R.T.Mead, R.L.McLean, F.Rudolph, N.W.Frenz: U.S. Patent No. 3994747 (1976)

4.43 W.Greatbatch: private communication (1977)

4.44 J.E.Oxley: A Solid State Electrochemical Capacitor, Extended Abstract No. 175, Electrochem. Soc. Meeting, Houston (1972) p. 446

4.45 J.A.Connelly: EDN **16**, 23 (1971)

4.46 Electronics **37** No. 29, 67 (1964)

4.47 Electronics **40** No. 7, 186 (1967)

4.48 J.H.Kennedy, F.Chen: J. Electrochem. Soc. **118**, 1043 (1971)

4.49 J.H.Kennedy, F.Chen, J.Hunter: J. Electrochem. Soc. **120**, 454 (1973)

4.50 H.Ikeda: Solid State Electrochemical Cell (Memoriode). In *Rechargeable Batteries in Japan*, ed. by Y.Miyake and A.Kozawa (JEC Press Inc., Cleveland 1977) Chap. 18, pp. 441—458

5. The β-Aluminas

J. H. Kennedy

With 13 Figures

The report by *Yao* and *Kummer* in 1967 [5.1] that β-alumina exhibited rapid sodium ion diffusion and high ionic conductivity has led to an intense effort to develop high energy batteries using β-alumina as a solid electrolyte. This effort has included research groups in France, Great Britain, Japan, the United States, and other countries. One particular battery being actively pursued operates at 300 °C with molten sodium as the anode and a molten sulfur/polysulfide mixture as the cathode material. Batteries operating at lower temperatures are also being developed. The β-alumina electrolyte is a critical feature in these batteries and must exhibit high ionic conductivity, negligible electronic conductivity, high mechanical strength, long life, and stability in corrosive environments. A large number of experimental techniques have been employed to increase our understanding of this interesting material and will be discussed in this chapter.

The history of β-alumina goes back at least 50 years before the work of *Yao* and *Kummer*. *Rankin* and *Merwin* [5.2] reported on the material which they thought was a crystallographic modification of alumina and adopted the designation β-alumina which has persisted to this day even though it is a misnomer. Later workers realized that sodium was always present, and by 1936 it was firmly established by *Ridgway* et al. [5.3] that β-alumina was actually a sodium aluminate. An early x-ray investigation carried out by *Hendricks* and *Pauling* in 1927 [5.4] determined the most likely space group to be $P6_3/mmc$ (D_{6h}^4), but it was not until the presence of alkali ions was included that the structure was finally established with the work of *Bragg* et al. [5.5], *Beevers* and *Brohult* [5.6], and *Beevers* and *Ross* [5.7].

5.1 Chemical Properties of β-Alumina

5.1.1 Formation and Fabrication

β-alumina is a general term that refers to a family of sodium aluminates with closely related structures and chemical properties. Materials prepared in the laboratory exhibit compositions in the range $Na_2O \cdot 5\,Al_2O_3$ to $Na_2O \cdot 9\,Al_2O_3$. The two most important members of the family are the hexagonal form (designated β) and the rhombohedral form (designated β″). The structural

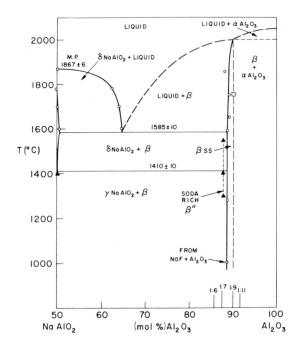

Fig. 5.1. Phase diagram of Al_2O_3–Na_2O system [5.8]

differences between β, β″, and other members will be discussed in Section 5.2. In addition, each member exhibits variable composition over a limited range leading to considerable confusion in the literature. The formula of β-alumina derived from the crystal structure is $Na_2O \cdot 11\,Al_2O_3$ but it always contains excess soda. The formula of β″-alumina is $Na_2O \cdot 5\,Al_2O_3$ but it is soda deficient and is normally stabilized by the addition of MgO and/or Li_2O [5.8].

The formation of β-alumina can be accomplished by heating Na_2CO_3 (or $NaNO_3$ or NaOH) with any of the Al_2O_3 modifications or hydrates to $\sim 1500\,°C$. Heating to $1100\,°C$ results in the formation of β″-alumina which can be converted to β-alumina and $NaAlO_2$ at $>1500\,°C$. β-alumina can also be prepared by reacting Na_2CO_3 with Al_2O_3 at $1000\,°C$ in the presence of NaF or AlF_3. However, it has not been possible to convert β″-alumina to β-alumina at $1000\,°C$ even in the presence of NaF so that the relative stability of these forms at temperatures $<1100\,°C$ is not known [5.8]. Both β- and β″-alumina exist indefinitely at room temperature at least in a dry atmosphere. The stability of β- and β″-alumina is summarized in the phase diagram determined by *Weber* and *Venero* [5.9] and shown in Fig. 5.1. Although it is not certain whether or not β-alumina melts congruently or not near $2000\,°C$, a eutectic melting point of $1585\,°C$ between β-alumina and $NaAlO_2$ is well established.

DeJonghe [5.10] carried out a study of β-alumina prepared by several laboratory procedures and compared its properties with β-alumina obtained commercially (Alcoa). The most satisfactory results were obtained by decomposing a mixture of $NaNO_3$ and $Al(NO_3)_3$ at $700\,°C$.

A detailed study of the phases produced during the synthesis of β- and β″-alumina was carried out by *Ray* and *Subbarao* [5.11]. The starting materials were Na_2CO_3, α-Al_2O_3, and MgO (for stabilizing β″-alumina and for doping β-alumina). In agreement with previous work it was found that β″-alumina decomposed at 1450 °C although MgO-stabilized β″-alumina was stable up to 1650 °C. A problem of segregation was noted, e.g., heating a Na_2CO_3:5 Al_2O_3 mixture to 1100 °C resulted in the formation of β″-alumina, α-Al_2O_3 and $NaAlO_2$ instead of pure β″-alumina. *Ray* and *Subbarao* postulated that at 851 °C, Na_2CO_3 melted and trickled to the bottom of the crucible making the bottom portions richer in soda (leading to $NaAlO_2$) and the top layers soda poor (leading to α-Al_2O_3).

The preparation conditions for a β-alumina sample are often tied to the fabrication conditions. Single crystals may be grown from a flux or melt while polycrystalline samples are usually fabricated by pressing powder at low temperature followed by sintering at 1600–1800 °C. In addition, hot-pressing techniques have been developed for β-alumina which are often carried out at ~1400 °C.

Early electrochemical studies used single crystals of β-alumina obtained by fracturing commercial fused cast bricks [5.1]. Later, single crystals of β- and MgO stabilized β″-alumina were grown using a flux evaporation technique at ~1600 °C [5.8]. A high soda composition of the Na_2O–Al_2O_3 mixture was used so that it was liquid (see Fig. 5.1) during the crystal growth operation. Platinum crucibles were used as containers and were covered with platinum foil. However, soda was lost slowly, leading to a sodium content appropriate to the formation of β-alumina.

More recently, advanced crystal growing techniques have been developed for the β-aluminas. *Morrison* et al. [5.12] and *Cocks* and *Stormont* [5.13] described an "edge-defined film-fed growth" method for growing ribbons and tubes. Because of the high sodium vapor pressure the growth chamber was maintained at $2 MN/m^2$ (>20 atm).

The Czochralski method for crystal growth was used by *Baughman* and *Lefever* [5.14] and patented by *Yancey* [5.15] for fabricating single crystals of β-alumina. In the Czochralski technique raw materials having a chemical composition corresponding to that of the desired crystal are melted in a refractory metal crucible. Then a single crystal seed of the desired crystal is dipped into the melt surface and slowly withdrawn. The liquid is cooled slightly as it is raised above the surface and crystallizes out on the tip of the seed crystal. *Baughman* and *Lefever* used a high soda melt composition ($Na_2O \cdot 4 Al_2O_3$) to eliminate the need for the high pressures used in the edge-defined film-fed growth method. *Yancey* was able to obtain pure β-alumina crystals at ambient pressures by adding 0.5–1.5% oxygen to the nitrogen atmosphere as opposed to the pure argon environment used by *Baughman* and *Lefever*.

Most studies of β-alumina have employed polycrystalline material instead of single crystals. Sintered compacts are easier to produce than growing large single crystals, and it is expected that practical applications will use polycrystalline material. On the other hand, it is essential to form near-theoretical density material to achieve high conductivity. The anisotropic nature of β-alumina leads to lower conductivity than properly oriented single crystals, and grain boundary effects are significant. In fact, the measured resistance of a β-alumina sample is often the result more of grain boundary resistance and electrode/electrolyte interface effects than the bulk electrolyte resistance. There are several published studies concerned with the production of high density sintered β-alumina (e.g., [5.10] and [5.16]) and innumerable patents describing techniques and special additives for achieving high quality material. Early work usually utilized sintering temperatures of 1700°–1850 °C to achieve high density, but more recent investigators have been able to sinter successfully at ~ 1600 °C by careful control of impurities such as Si and Ca and with the use of a sintering aid [5.10]. The additive consists of a sodium-aluminum oxide mixture (Na/Al ~ 0.54) corresponding to the eutectic composition melting at 1585 °C (Fig. 5.1).

Fabrication temperatures for polycrystalline material have been lowered even more with the advent of hot-pressing techniques. *Clendenen* and *Olson* [5.17] patented techniques for hot-pressing, and more recent work has been carried out by *Virkar* et al. [5.18] and by *McDonough* and *Rice* [5.19]. Whereas β-alumina is the normal product in high temperature sintering, β″-alumina can be the major product during hot-pressing because the temperature is ~ 1400 °C, the region in which β″-alumina is stable. *Virkar* et al. [5.18] showed that after hot-pressing for 15 min at 1350°–1400 °C under 4500 psi pressure, a significant amount of β-alumina was present, but this could be converted to β″-alumina by annealing for several hours at 1400 °C in a sealed container. Hot-pressing resulted in some preferred orientation with the β″-alumina grains aligned with many of their *c* axes coincident with the pressing direction. The effect resulted in only a 25 % increase in resistance in this direction compared with the resistance in the plane of the hot-pressed disc.

Other fabrication techniques have been applied to produce specific shaped pieces of β-alumina. One in particular which has received considerable attention is electrophoresis in which the β-alumina powder is suspended in an organic liquid and migrates to an electrode when an electric field is applied [5.20–24]. Several mechanisms have been proposed for generating charged particles of β-alumina including incorporation of H^+ from an acidic solvent [5.20a] and loss of Na^+ from the β-alumina [5.21]. The actual process may be more complicated than either of these two simple mechanisms [5.23, 24]. Plasma spray techniques have also been investigated by *Markin* et al. [5.25] to fabricate β-alumina tubes. Spheres of β-alumina (15 μm diameter) needed for plasma spraying were produced by calcining an Al_2O_3 gel doped with Na^+ at 1000 °C.

5.1.2 Reactivity

One of the reasons that β-alumina has a bright future as a solid electrolyte material is its lack of reactivity. The material is ceramic in nature and is stable at elevated temperatures even in the presence of molten sodium. The thermodynamic stability of β-alumina is essentially the same as $Al_2O_3 + Na_2O$. Based on the stoichiometric formula of $Na_2Al_{22}O_{34}$, *Kummer* [5.8] tabulated free energy of formation as a function of temperature from data obtained by *Weber*. From these values the free energy change for the formation of β-alumina from Al_2O_3 and Na_2O is calculated to be -53 kcal/mole of β-alumina at 1073 °C. *Kummer* also calculated ΔG^0 for the reaction

$$2\,NaAlO_2 + 10\,Al_2O_3 \rightarrow Na_2Al_{22}O_{34} \qquad (5.1)$$

to be -9.0 kcal/mole of β-alumina at 1000 °C.

One reaction of β-alumina which causes concern at high temperatures is the loss of Na_2O, especially during sintering at > 1600 °C. Many investigators sinter β-alumina in a bed of coarse β-alumina to maintain the equilibrium Na_2O vapor pressure, and even under these conditions it is important to analyze the final product for sodium content because a variation of several percent is not uncommon [5.26]. The rate of vaporization of alkali oxide from Na^+-β-alumina and K^+-β-alumina (see Sect. 5.3) was measured by *Sasamoto* and *Sata* [5.27]. Vaporization rates of 10^{-7}–10^{-8} g/cm^2–s (Knudsen Cell method) were observed for nonporous sintered material in the temperature range 1200–1300 °C. Species in the vapor phase identified by mass spectrometry included Na, K, O, and O_2, and it appeared that the vaporization rate was controlled by a decomposition reaction.

An important reaction of β-alumina at room temperature is with water and H_3O^+. Na^+-β-alumina can be ion exchanged with other cations (see Sect. 5.3) and, therefore, it is not surprising that sodium can be replaced by hydronium ions in boiling concentrated H_2SO_4 [5.28]. However, nmr results on β-alumina powder exposed to moist air indicated the presence of both H_2O and mobile protons [5.29]. The water was desorbed at 230 °C in a dry N_2 atmosphere. *Will* [5.30] studied the occlusion of water and its effect on the conductivity of β-alumina and showed that there was a rapid exothermic occlusion of water, probably by micropores and cleavage planes, reaching saturation in <1 h. Water-induced cleavage of lattice planes was observed by *Will* and others on β-alumina single crystals. The presence of H_2O in the planes affected the high frequency conductivity only slightly while the exchange of Na^+ by H_3O^+ led to significant increases in resistivity. The presence of H_2O caused a dramatic increase in grain boundary effects as evidenced by the low DC conductivity compared to dry β-alumina. Water could be oxidized electrochemically at a Pt/β-alumina interface with H^+ entering the β-alumina [5.31].

β-alumina may also be somewhat unstable in the presence of base. *Yamaguchi* and *Kato* [5.32] showed that the following reaction took place above 70 °C

$$Na_2O \cdot 11\,Al_2O_3 + XNa_2CO_3 \rightarrow (1+X)Na_2O \cdot 11\,Al_2O_3 + XCO_2\uparrow. \qquad (5.2)$$

The most Na_2O-rich β-alumina formed was $1.9\,Na_2O \cdot 11\,Al_2O_3$ when the reaction was carried out below 450 °C. Above 620 °C, β-alumina reacted with Na_2CO_3 to form $NaAlO_2$.

5.2 Structure of β-Alumina

5.2.1 X-Ray Diffraction

Most of the information concerning the structure of β-alumina has been obtained using x-ray diffraction, although in recent years several other experimental probes have been applied to determine details of the structure and conductivity mechanism. It was mentioned earlier that the basic crystal structure was established by *Beevers* and *Ross* [5.7] in 1937. The structure they determined is shown in Fig. 5.2. The blocks of Al^{3+} and O^{2-} are packed in the same fashion as the packing in spinel, $MgAl_2O_4$, and are usually called "spinel blocks". In this case, Al^{3+} ions occupy the octahedral sites as well as the tetrahedral sites occupied by Mg^{2+} ions in spinel. The spinel-type blocks are separated by a loosely packed plane containing Na^+ and O^{2-}. Because of the loose packing, space is available for movement of the sodium ions leading to the high ionic conductivity shown by β-alumina. However, conductivity is limited to this plane and movement along the c axis is exceedingly difficult. The material, therefore, is highly anisotropic.

The conduction plane of β-alumina is a mirror plane, and thus, the face-centered cubic packing arrangement of oxide ions "reflects" from this mirror plane as shown in Fig. 5.3. This packing arrangement is slightly different in β″-alumina since the conduction plane is not a mirror plane. As can be seen in Fig. 5.3, it takes 3 spinel-type blocks before the stacking arrangement is repeated, and for this reason, β″-alumina is called "3-block" material while β-alumina is called "2-block".

Although spinel is cubic, the additional conduction planes lead to a hexagonal crystal structure for β-alumina and a rhombohedral structure for β″-alumina. The lattice constants for β-alumina are $a = 5.59$ Å and $c = 22.53$ Å. The structure and lattice constants for β″-alumina were determined by *Yamaguchi* and *Suzuki* [5.33, 34] to be $a = 5.59$ Å and $c = 34.23$ Å (potassium form). A more complete x-ray crystallographic investigation was carried out by *Bettman* and *Peters* [5.35] who found $a = 5.614$ Å and $c = 33.85$ Å for the sodium form of β″-alumina stabilized by magnesium. This compound contains considerably more sodium than β-alumina and has the idealized formula $Na_2MgAl_{10}O_{17}$.

Other modifications of the spinel block stacking arrangement have been reported and given the names β‴-alumina [5.36] and β⁗-alumina [5.36]. These modifications contain six oxide ion layers in each spinel-type block and therefore the c axis is larger than in β-alumina. β‴-alumina is a 2-block material with $a = 5.62$ Å and $c = 31.8$ Å. β⁗-alumina is a 3-block material and bears the same structural relationship to β‴-alumina as β″-alumina does to β-alumina. β′-alumina is a variation discussed by *Yamaguchi* and *Suzuki* [5.34] but has not been verified by other workers. It was probably β-alumina with a high sodium concentration. Because all these materials exhibit variable composition it is often difficult to decide when the limit of one form has been reached and a new one begins. The regions of stability for the various β-aluminas are shown in Fig. 5.4.

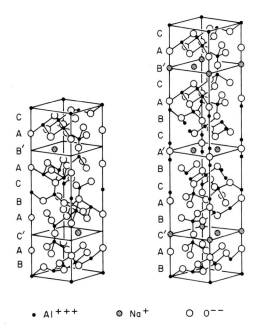

Fig. 5.2. Structure of β-alumina (left) and β″-alumina (right). The Na atoms in β-alumina are shown in Beevers-Ross (BR) sites

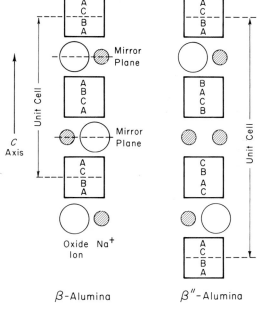

β-Alumina

β″-Alumina

Fig. 5.3. Oxide ion packing arrangement in β-alumina and β″-alumina (Note: letters refer to stacking arrangement where ABC represents face-centered cubic packing while ABAB would represent hexagonal packing)

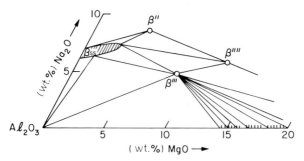

Fig. 5.4. Regions of stability for various β-aluminas at 1700 °C. Lines in the lower right represent tie lines in the β‴-spinel solid solution field. A line joins an analyzed composition with the composition of spinel in equilibrium as inferred from the x-ray lattice constant of the spinel. The β-alumina solid solution field (β_{ss}) was mapped by noting the disappearance of other phases [5.8]

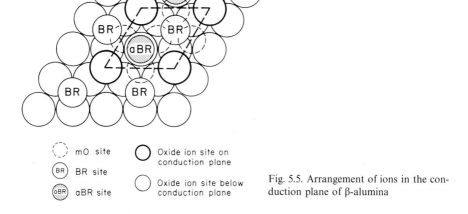

Fig. 5.5. Arrangement of ions in the conduction plane of β-alumina

Beevers and *Ross* [5.7] determined the most probable position for the sodium ion in β-alumina, and this is shown in Fig. 5.5. The sites labeled BR refer to these "Beevers-Ross" positions. All these sites would ordinarily be filled in the stoichiometric $NaAl_{11}O_{17}$ material. However, *Felsche* [5.37] found that the sodium sites were only partially occupied using three-dimensional refinement methods. *Peters* et al. [5.38] studied typical crystals containing 29% excess sodium and concluded that the sodium was smeared out in the conduction plane. They postulated that the excess sodium was charge compensated by aluminum vacancies so that the formula could be written more accurately as $Na_{1+x}Al_{11-x/3}O_{17}$ where X is usually 0.15–0.30. Two possible positions for the excess sodium ions are shown in Fig. 5.5. The sites labeled aBR refer to "anti-Beevers-Ross" positions since *Beevers* and *Ross* rejected this type of site. The other positions lie between the oxide ions and are labeled mO for "mid-oxygen".

Peters et al. [5.38] measured the electron density due to Na^+ and found that the BR sites were only 75% occupied. No electron density due to Na^+ was found at aBR sites and the remaining Na^+ electron density was found in a diffuse fashion around the mO sites.

Roth used Ag^+-β-alumina in place of the sodium form to take advantage of the large silver atomic form factor [5.39]. Least squares and Fourier analysis showed that about 27% excess silver was present and that all the silver ions were in the conduction plane. However, it appeared from density measurements that charge compensation was accomplished with the presence of interstitial oxide ions leading to the formula $Ag_{1.27}Al_{11}O_{17.135}$. Another dramatic difference was the fact that occupation of aBR sites was about 43%. Occupation of the BR sites was about 84% with little or no Ag^+ in mO sites (reader should note that uncertainties in these occupancy figures are probably $\pm 10\%$). Thus, it appears from these two studies that sodium ions and silver ions occupy different sites in β-alumina in addition to both residing in the BR sites. This conclusion has been supported by other experimental probes which will be discussed later. The percentage occupations have been quoted at room temperature, and it should be kept in mind that there is a temperature dependence. There is evidence [5.40] that as the temperature increases the structure becomes increasingly disordered until the distribution approaches that of a two-dimensional liquid.

A similar single crystal x-ray diffraction study of K^+-β-alumina and cobalt-doped K^+-β-alumina was carried out by *Dernier* and *Remeika* [5.41]. They showed that potassium behaved in a fashion similar to sodium in occupying BR and mO sites with no occupation of aBR sites. They also found that cobalt resided exclusively in a tetrahedral site normally occupied by aluminum. However, for the K^+-β-alumina this tetrahedral site was probably under-occupied and may be reflecting an aluminum ion deficiency to account for charge compensation.

The Tl^+-β-alumina analog was studied by *Kodama* and *Muto* [5.42] who found electron density due to the presence of Tl^+ at BR and aBR positions with no density at mO sites similar to Ag^+-β-alumina. Percentage occupation of the various sites for Na^+-, K^+-, Ag^+-, and Tl^+-β-alumina is summarized in Table 5.1. It should be noted that there is a large uncertainty in these values, and they most likely vary with the total concentration of M^+.

At this point one might ask why Na^+ and K^+ occupy different sites in β-alumina compared to Ag^+ and Tl^+. An examination of the size and coordination number of these sites provides some clues. The BR site has octahedral coordination to the neighboring oxide ions in the spinel-type blocks above and below the conductive plane. The M–O distance is 2.87 Å which is sufficiently large (M^+ radius $\leq 2.87 - 1.40 = 1.47$ Å) for all M^+ ions up to Rb^+. Therefore, it is understandable that most of the mobile ions reside in BR positions. The aBR site lies directly between oxide ions in the spinel-type blocks which means the coordination number is two, and the M–O distance is only 2.38 Å (the distances will vary a little depending on the particular mobile ion present). This is sufficiently large for Na^+ and Ag^+ but too small for most M^+ ions. The low

Table 5.1. Occupation of various sites by mobile ions in β-alumina

Compound	Relative percentage occupation in site[a]			Ref.
	BR	aBR	mO	
Na$^+$-β-alumina	59	0	41	[5.38, 42]
K$^+$-β-alumina	54	0	46	[5.41]
Ag$^+$-β-alumina	66	34	0	[5.39]
	53	34	13	[5.42]
Tl$^+$-β-alumina	70	30	0	[5.42]

[a] Relative percentage occupation $=(M^+$ electron density at specific site/Total M^+ electron density)$\cdot 100$. By using this definition, occupation percentages have been normalized to the stoichiometric formula. Absolute percentage occupation can be found by multiplying the relative value by the concentration, e.g., for Na$^+$-β-alumina containing 29% excess Na$^+$ the BR site would be $0.59 \cdot 1.29 = 0.76$ (76%) occupied. However, it should be noted that there are three mO sites for every BR and aBR site so that even with 41% of the Na$^+$ in mO sites they would be only $0.41 \cdot 1.29/3 = 0.18$ (18%) occupied for Na$^+$-β-alumina containing 29% excess Na$^+$.

coordination number is probably the factor which accounts for the tendency of Ag$^+$ to occupy aBR sites while Na$^+$ does not. Thallium ions appear to reside slightly away from the aBR site, and this position was designated aBR' by *Kodama* and *Muto* [5.42]. The size is somewhat larger, allowing the highly polarizable Tl$^+$ to "squeeze" in. The mO positions again have a high coordination number (8 including the two neighboring oxides on the conduction plane) and are only slightly smaller than the BR sites (~ 2.8 Å). The alkali ions appear to prefer the high coordination number sites while the more covalent, highly polarizable ions partially occupy low coordination number sites.

5.2.2 Neutron Diffraction

Several other experimental probes have been used to gather information concerning the structure of β-alumina especially in regard to the sites occupied by mobile ions in the conduction plane. As was mentioned in the previous subsection Ag$^+$-β-alumina was used in x-ray studies because of its large structure factor. Neutron diffraction is better suited to the study of light elements, and, therefore, it has been used for the study of Na$^+$-β-alumina.

Roth et al. [5.40, 43, 44] studied β- and β″-alumina (stabilized by Mg) using a three-dimensional neutron diffraction analysis. They showed that Na$^+$ was distributed in two equivalent sites in β″-alumina. It should be noted that the conduction plane in β″-alumina is not a mirror plane, which results in slightly different site geometry and size distribution compared to β-alumina. It was also shown that interstitial oxide ions resided on the conduction plane in β-alumina which would act to impede the mobility of Na$^+$. The authors concluded that these interstitial oxide ions could account for charge compensation. No blocking oxide ions were found in β″-alumina (charge compensation is accomplished by the stabilizing Mg^{2+} in the spinel block) and this helps to account for the higher conductivity found for β″-alumina. The difference Fouriers also showed that the magnesium in β″-alumina substituted for aluminum in only one of the four sets of crystallographic sites in the spinel block. This site has tetrahedral coordination and was the same site occupied by Co^{2+} described previously [5.41].

5.2.3 Neutron and X-Ray Diffuse Scattering

X-ray diffuse scattering is a tool well suited for the study of partially ordered materials and gives an insight on local order of atoms which can range continuously from a slightly disordered crystal to an isotropic amorphous (liquid-like) state. *Boilot* et al. [5.45] and *Le Cars* et al. [5.46] found that Ag^+-β-alumina was quasi-liquid at 750 K while at 20 K the Ag^+ ions formed quasicrystalline short range ordered domains. The order and size of these domains varied with temperature and with the specific mobile ion in the structure (Ag^+, Na^+, K^+, Tl^+).

McWhan et al. [5.47] carried out x-ray diffuse scattering measurements on β-alumina containing 26% excess mobile ion (Na^+, Li^+, K^+, Rb^+, Ag^+, and Eu^{2+}). The modulation of the diffuse scattering in the [001] direction was related to the nature of the compensating defect and the correlations between defects and diffusing ions. Evidence again pointed to interstitial oxide ions blocking the mobile ions, with a strong correlation between the ionic conductivity and the degree of order between the compensating defects.

McWhan et al. [5.48] also used neutron scattering techniques to measure elastic constants of Na^+-, Ag^+-, K^+-, and Rb^+-β-alumina. It was found that β-alumina was "softer" than Al_2O_3 and that the transverse acoustic mode propagating along the c axis was sensitive to mobile ion substitution. The associated elastic constant (C_{44}) increased by about 50% when Rb^+ was substituted for Na^+. This was consistent with the idea that with increasing size of the diffusing ion the local strain fields increase and that this in turn increases the ion-ion interactions, i.e., an increased binding between the diffusing ions and the spinel blocks. Additional neutron scattering studies have been carried out by *Shapiro* [5.49] and *Tofield* et al. [5.50] showing the dynamical behavior of the mobile ions and their interaction with other ions in the solid.

5.2.4 Nuclear Magnetic Resonance

The earliest report of using nmr to study β-alumina is that of *Kline* et al. [5.29]. Nuclear magnetic relaxation experiments provide a means to study the motion of a nuclear probe (^{23}Na, $I = 3/2$) in a crystal. The room temperature nmr spectrum for ^{23}Na in polycrystalline β-alumina consisted of two well resolved absorption peaks. The dipolar linewidths showed motional narrowing with increasing temperature in the range $-170°$ to $+250°C$. The narrowing phenomenon was most pronounced from $-100°$ to $25°C$ where the linewidth decreased from 6 to 3 G. By estimating a rigid lattice bandwidth of 10–15 G, *Kline* et al. were able to calculate a diffusional jump frequency as a function of temperature. An Arrhenius plot of the results gave an activation energy of ~ 0.1 eV, in reasonable agreement with the activation energy for Na^+ diffusion (0.17 eV). No motional narrowing was found for ^{27}Al as expected. The effect of water on β-alumina was observed from both ^1H nmr and changes which occurred in the ^{23}Na nmr. Approximately 1.8 moles H_2O/mole β-alumina were absorbed by the fine powder which was subsequently lost in three stages when the β-alumina was heated in dry N_2: I) 75 % was lost at 100–150 °C, II) 7 % was lost at 150–200 °C, and III) the remainder lost gradually between 200–500 °C. The ^{23}Na quadrupole interaction decreased markedly with increasing water content.

Jérome and *Boilot* [5.51] measured spin-lattice relaxation times of ^{23}Na between 120 and 800 K and found evidence for two relaxation times perhaps indicative of two types of Na^+ motion. Arrhenius plots of these relaxation times gave activation energies of 0.2 and 0.1 eV. The authors hypothesized that they might represent different motions involving Na^+ ions residing on BR and mO sites.

Chung et al. [5.52, 53] carried out ^{23}Na nmr measurements using single crystals of β-alumina and found an activation energy for sodium ion motion of 0.17 eV, in good agreement with diffusion experiments. In a more recent study *Bailey* et al. [5.54] gave evidence for a number of independent sodium ion sites at low temperature (77–103 K) and attributed this to a distribution of Na^+ among BR, mO, and aBR sites. Above 110 K the structure gave way to a single, symmetric resonance line indicative of rapid ionic motion among lattice sites related by threefold symmetry. Activation energy in the higher temperature region was 0.1 eV, somewhat lower than the diffusion value and the value of *Chung* et al. [5.53]. *Walstedt* et al. [5.55] also reported activation energies from nmr data of ~ 0.1 eV for polycrystalline material but only ~ 0.04 eV for single crystals.

5.2.5 Infrared and Raman Spectroscopy

The earliest infrared spectra of β-alumina were used to investigate the incorporation of water and H^+ substitution in the conduction plane [5.8]. Single crystals of Na^+-β-alumina exposed to ambient air for several hours exhibited an O–H band near 3100 cm^{-1} and this band became larger when the sample was heated to 300 °C for 2 days in 25 mm H_2O vapor. The results were attributed to incorporation of water molecules because H^+-β-alumina (prepared by treating

Ag$^+$-β-alumina with H$_2$ at $\geqq 300\,°$C) gave peaks at ~ 3500 cm^{-1} [5.8]. Exchange of hydrogen for deuterium moved the absorption band to 2600 cm^{-1} [5.8].

Far infrared spectra from 5–400 cm^{-1} were reported by *Armstrong* et al. [5.56] for Na$^+$-β-, Ag$^+$-β-, and Na$^+$-β″-alumina. Vibrations due to mobile ions were noted and compared with ir spectra (400–1200 cm^{-1}) of α-Al$_2$O$_3$ which contains no mobile ions.

Risen and *Butler* [5.57] assigned far infrared bands to vibrations at the BR, aBR, and mO sites by using a site symmetry interpretation. For Na$^+$- and K$^+$-β-alumina BR and mO vibrations were found while Ag$^+$- and Tl$^+$-β-alumina exhibited vibrations attributed to ions occupying BR, aBR, and mO sites. Raman spectra were also reported.

One of the questions which has been posed concerning the mechanism of conduction for ions in β-alumina is whether the ion moves by a diffusion process consisting of discrete jumps from one site to another with a characteristic jump frequency, or whether it is in a free-particle state characterized by a mean free path. Infrared spectra have been interpreted according to the diffusion model [5.58]. Raman scattering experiments carried out by *Chase* et al. [5.59] on Na$^+$- and Ag$^+$-β-alumina also implied a diffusion model throughout the temperature range of 4.2 to 900 K. Raman peaks were observed at 60 cm^{-1} for Na$^+$-β-alumina and at 25 cm^{-1} for Ag$^+$-β-alumina. The intensity of the Ag peak was about 50 times that of Na, probably indicative of the higher polarizability of the silver ion. The frequencies correlated with the inverse square roots of their masses (i.e., $v_{Ag}\sqrt{M_{Ag}} = v_{Na}\sqrt{M_{Na}}$) and these vibrational characteristics changed little with temperature. The only significant difference between Raman and infrared data was that the Raman absorption peak for Na$^+$-β-alumina was much narrower than the infrared peak.

5.2.6 Electron Microscopy

The wide spacing between conduction planes (11.3 Å) has made it possible to resolve them in the electron microscope [5.60–63]. In addition to the spectacular aspect of "seeing" this interesting structure directly, information has been obtained showing lattice bending and rotation [5.60] and disordered intergrowths [5.61]. *De Jonghe* [5.62] used scanning electron microscopy to reveal networks of planar faults leading to an alteration of the conduction plane at the intersection of faults. Low-angle tilt boundaries observed were suggested to cause inhomogeneity in sodium ion current flow. Depending on the dislocation resistance value, the tilt boundary could either increase or decrease current density in this region and possibly lead to electrolyte breakdown [5.62b]. The possibility of observing the coexistence of various β structures and the role that additives such as Mg play in affecting the detailed structure will aid in our understanding of β-alumina subtleties.

5.3 Conductivity and Diffusion in β-Alumina

5.3.1 Ion Exchange Properties

One of the properties of β-alumina which was recognized early [5.64, 65] and indicated that mobile ions were present was the ability to replace sodium ions by other metal cations using ion exchange techniques. Because water itself may enter the conduction plane [5.29, 30], exchange experiments were normally carried out using molten salts. A classic study was carried out by *Yao* and *Kummer* [5.1] in which sodium ions were exchanged in molten salts (300–350 °C) by Ag^+, K^+, Rb^+, Li^+, NH_4^+, and Cs^+ (partial). In general, the mole fractions of Na^+ in the melt and in the solid β-alumina at equilibrium were different as shown in Fig. 5.6. The shape of these curves will be determined by the ratio of the activities of the two salts in the melt and the ratio of the activities of the two cations in the solid.

In addition to the monovalent ions reported by *Yao* and *Kummer* listed above, complete exchange for Na^+ in β-alumina has been reported for In^+ via Ag^+-β-alumina [5.1], NO^+ [5.66], Ga^+ [5.67], Cu^+ [5.68], and H_3O^+ [5.28]. Exchange experiments were also reported by *Yao* and *Kummer* [5.1] using divalent ions, and some exchange was noted with Sr^{2+}, Pb^{2+}, Fe^{2+}, Ba^{2+}, Sn^{2+}, Mn^{2+}, and Ca^{2+}. However, in a number of these cases, the single crystals fractured, which was attributed to a shortening of the a-axis with M^{2+} exchange. As expected from charge balance considerations, one divalent ion replaced two monovalent ions. Ion exchange with monovalent ions also produced lattice constant changes as summarized in Table 5.2 [5.8]. Little change was noted in the a-axis, but the c-axis increased in general as the monovalent ion increased in

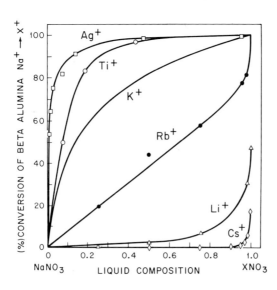

Fig. 5.6. Ion exchange equilibria between β-alumina and binary nitrate melts containing $NaNO_3$ and another metal nitrate at 300–350 °C [5.1]

Table 5.2. Lattice constants of exchanged β-aluminas

Ion	Ionic radius[a] [Å]	a^b [Å]	c [Å]	Δc [Å]
Na^{+c}	0.95	5.594	22.530	0
Li^+	0.60	5.596	22.570	+0.040
Ag^+	1.26	5.594	22.498	−0.032
Ga^{+d}	—	5.600	22.718	+0.188
K^+	1.33	5.596	22.729	+0.199
NH_4^+	1.43	5.596	22.888	+0.358
Tl^+	1.44	5.598	22.93	+0.40
Rb^+	1.48	5.597	22.883	+0.347
H_3O^{+e}	—	—	—	+0.125
H^+	—	5.602	22.677	+0.147
NO^+	—	5.597	22.711	+0.181

[a] Ionic radii according to Pauling.
[b] Based on K_{α_1} for copper of 1.54040 Å.
[c] The β-alumina crystals contained ∼25% excess soda.
[d] Ref. [5.42].
[e] The absolute values for the a- und c-axis determined by *Saalfeld* et al. [5.28] have been omitted because their value of Na$^+$-β-alumina was larger than usual.

size. It is reasonable to assume that the observed expansion of the c-axis occurred in the conduction plane containing the monovalent ions.

In some cases direct exchange did not take place, e.g., *Yao* and *Kummer* [5.1] reported that treatment of 1 mm crystals of Na$^+$-β-alumina with molten InI at 400 °C produced no measurable exchange. However, In$^+$-β-alumina could be formed by treating Ag$^+$-β-alumina with In metal. The following reaction took place

$$In + Ag^+\text{-}\beta\text{-alumina} \rightarrow Ag + In^+\text{-}\beta\text{-alumina} \,. \tag{5.3}$$

This redox exchange method was also used to produce H$^+$-β-alumina [5.8]. Indirect methods were used by *Nishikawa* et al. [5.69] to avoid problems encountered with aqueous solutions [5.29, 30]. First Na$^+$-β-alumina was converted to H$_3$O$^+$-β-alumina by treating 325 mesh powder with concentrated H$_2$SO$_4$ at 240 °C for 20 min. Then the H$_3$O$^+$ could be replaced by K$^+$ or Li$^+$ by immersion in aqueous solutions saturated with KOH or LiOH for 2 h at 240 °C.

5.3.2 Tracer Diffusion Measurements

The fact that ion exchange took place rapidly at ∼300 °C indicated that the monovalent ions in the conduction plane were mobile and should, therefore, exhibit high diffusion coefficients. *Yao* and *Kummer* [5.1] measured self-diffusion coefficients for Na$^+$, K$^+$, Rb$^+$, and Ag$^+$ in β-alumina crystals by employing the radioactive isotopes ^{24}Na, ^{42}K, ^{86}Rb, and ^{110}Ag produced by neutron irradiation. After activation the irradiated crystals had to be annealed 1–2 h at 1100 °C because many of the ions had been pushed into nonexchangeable lattice positions. The annealed crystals were placed, for various intervals of time, in a molten salt containing the same ion as the crystal and the reduction of radioactivity of the crystal was measured. The exchange rate was determined only by the dimensions perpendicular to the c-axis indicating that two-dimensional diffusion took place in the conduction plane with little or no diffusion along the c-axis. The calculation of D, the diffusion coefficient, by this method involved some geometric approximation, because the crystals were not rectangular or circular, but was probably accurate to ±5%. Diffusion constants for the equation

$$D = D_0 e^{-E_a/RT} \quad (200–400 \,^\circ C) \tag{5.4}$$

are summarized in Table 5.3.

Yao and Kummer [5.1] used the stable isotope ^{6}Li for the diffusion measurements reported in Table 5.3 for lithium. ^{6}Li$^+$-β-alumina was prepared from Ag$^+$-β-alumina and contacted with ^{7}LiNO$_3$ melts for various times. The residual ^{6}Li content was found by neutron irradiation of the single crystals in the reactor. The series of nuclear reactions which took place led to ^{18}F which was then counted.

Thallium diffusion in β-alumina was measured by Radzilowski et al. [5.70] using ^{204}Tl as the isotopic tracer.

Diffusion of ^{22}Na in polycrystalline β-alumina has been measured by Dunbar and Sarian [5.71]. They used a sectioning technique and measured activity as a function of penetration distance by the radioactive isotope at a particular time t. The activity as a function of distance was given by the equation

$$C^*(x, t) = [A/(\pi D^* t)^{1/2}] \exp(-x^2/4D^* t). \tag{5.5}$$

By plotting log (activity) vs x^2, a straight line was obtained with a slope of $-1/4D^* t$. The diffusion coefficient was $3.5 \cdot 10^{-6}$ cm^2/s at 400 °C, and other data are reported in Table 5.3. The material used was hot-pressed and showed a higher diffusion coefficient when measured perpendicular to the pressing direction, indicating less orientation along the c-axis in this direction. Terai et al. [5.72] obtained similar results with hot-pressed polycrystalline β-alumina.

5.3.3 Conductivity of Single Crystal β-Alumina

There are two basic approaches to the measurement of the conductivity of solid electrolytes. Conceptually the most straightforward is to use reversible electrodes and a DC signal with the resistance calculated from Ohm's Law. In the case of β-alumina this would require some type of sodium electrode. The second approach is to use irreversible electrodes and an AC signal. Bridge techniques are used to ascertain the impedance of the sample, and, by noting frequency

Table 5.3. Comparison of activation energies for tracer diffusion with activation energies for conductivity (single crystal samples)

Ion	D_0	E_a [kcal/mole]		Calc. D_t(25°C)	$\sigma_{25°C}$	Calc. D_σ(25°C)	D_t/D_σ
	[cm²/s]	D	σ	[cm²/s]	[(Ω cm)⁻¹]	[cm²/s]	
Na⁺	$2.4 \cdot 10^{-4}$	3.81	3.78	$4.0 \cdot 10^{-7}$	$140 \cdot 10^{-4}$	$6.8 \cdot 10^{-7}$	0.59
Na⁺ᵃ	$2.0 \cdot 10^{-4}$	3.60	—	$5 \cdot 10^{-7}$	—	—	—
Ag⁺	$1.65 \cdot 10^{-4}$	4.05	3.98	$1.7 \cdot 10^{-7}$	$64 \cdot 10^{-4}$	$3.1 \cdot 10^{-7}$	0.55
K⁺	$0.78 \cdot 10^{-4}$	5.36	6.78	$9.6 \cdot 10^{-9}$	$0.65 \cdot 10^{-4}$	$3.17 \cdot 10^{-9}$	3.0(?)
Rb⁺	$0.34 \cdot 10^{-4}$	7.18	—	$1.9 \cdot 10^{-10}$	—	—	—
Tl⁺	$0.65 \cdot 10^{-4}$	8.22	8.19	$6.2 \cdot 10^{-11}$	$0.02 \cdot 10^{-4}$	$9.76 \cdot 10^{-11}$	0.63
Li⁺	$14.5 \cdot 10^{-4}$	8.71	8.54 high temp 4.30 low temp	—	$1.3 \cdot 10^{-4}$	—	—

ᵃ Hot-pressed polycrystalline sample [5.71].

dependence, the electrolyte resistance can often be separated from various capacitance effects (geometric, double layer, and grain boundary for polycrystalline samples).

   Imai and *Harata* [5.73] measured the conductivity of single crystal β-alumina using reversible molten sodium electrodes and reported 0.0765 $(\Omega \,\text{cm})^{-1}$ at 100°C just above the melting point of sodium. The conductivity at room temperature was 0.033 $(\Omega \,\text{cm})^{-1}$ found by extrapolation.

   A novel approach to the problem of sodium electrode preparation was taken by *Whittingham* and *Huggins* [5.74] who used tungsten bronze electrodes. Tungsten bronze, $Na_x WO_3$, exhibits both ionic and electronic conductivity and functioned as a reversible sodium electrode in cells of the type

$$Na_x WO_3/\beta\text{-alumina}/Na_x WO_3 .$$

The conductivity of β-alumina at room temperature was 0.014 $(\Omega \,\text{cm})^{-1}$, and measurements covered the range $-150°C$ to $800°C$ as shown in Fig. 5.7. The difference between the data of *Whittingham* and *Huggins* and those of *Imai* and *Harata* may represent a difference in sample composition because the conductivity is sensitive to sodium content. However, it may also reflect the difference in electrodes and the necessary extrapolation from 100°C in the case of the study by *Imai* and *Harata*.

   Conductivity measurements using indium irreversible ("blocking") electrodes were made by *Yao* and *Kummer* [5.1] who reported a value of 0.033 $(\Omega \,\text{cm})^{-1}$ in the frequency range of 0.5–1.5 MHz. This value agrees well with the molten sodium electrode technique although it should be noted that because of the small size of the single crystals and the necessity for very high frequencies, the measurements by *Yao* and *Kummer* were not highly accurate.

   More recent measurements on single crystal β-alumina ($Na_2O \cdot 8Al_2O_3$ composition) were carried out by *Fielder* et al. [5.75] on cylindrical boules produced by Union Carbide Corp. AC measurements with sputtered Pt

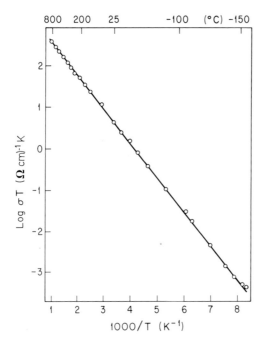

Fig. 5.7. Conductivity of single crystal β-alumina as a function of temperature [5.74]

electrodes gave a value of 0.03 $(\Omega \, cm)^{-1}$ at 1 MHz. A 4-probe DC technique (voltage drop across two middle electrodes used in Ohm's Law calculation eliminates electrode/electrolyte polarization effects occurring at outer electrodes) gave a value of 0.024 $(\Omega \, cm)^{-1}$ at 25 °C extrapolated from data at 40 °C.

As one might expect, the conductivity of β-alumina can change greatly when sodium is replaced by another mobile ion. Few studies have been carried out with these materials, but *Whittingham* and *Huggins* [5.76] did measure single crystal β-alumina in the Na^+-, K^+-, Tl^+-, and Li^+-forms using bronze-type electrodes as described above [5.74]. Conductivity and activation energies for conduction are given in Table 5.3. A detailed study of Ag^+-β-alumina was also carried out by *Whittingham* and *Huggins* [5.76b] using reversible silver electrodes. Results are shown in Table 5.3, and it should be noted that the Arrhenius plot was linear from 23 °C to nearly 800 °C with an activation energy of 3.98 kcal/mole. The close agreement between activation energies for diffusion and conductivity given in Table 5.3 indicates that the mechanism for these two transport processes must be the same. The relationship between conduction and diffusion will be discussed more fully in a later section. It can also be seen from Table 5.3 why most research has been carried out using Na^+-β-alumina. Aside from Ag^+, all other M^+-β-aluminas exhibit conductivities more than two orders of magnitude smaller.

The fact that β-alumina in the Na^+ form exhibits the highest conductivity probably indicates that the Na^+ is the ideal size. Larger ions may experience large electrostatic repulsion as they move from site to site leading to high activation energies and lower conductivities. Lithium ions may be small enough to reside near oxide ions instead of riding through the middle of the conduction plane channel and thus again must overcome a higher barrier than Na^+ to move from site to site. If this hypothesis is true, an increase in pressure would decrease the conductivity of larger ions but could increase the conductivity of Li^+-β-alumina by forcing it to ride more nearly in the center of the conduction plane. *Radzilowski* and *Kummer* [5.77] measured the resistance of K^+-β-alumina, Na^+-β-alumina, and Li^+-β-alumina and found that indeed the resistance increased with applied pressure (up to 4 kbar) for K^+-, decreased for Li^+- and was nearly constant for Na^+-β-alumina. At higher pressures (to 100 kbar) *Itoh* et al. [5.78] found that the resistance of Na^+-β-alumina increased about fivefold from one atmosphere.

The conductivity of β″-alumina is greater than that of β-alumina as can be seen in Fig. 5.8 [5.8]. However, conductivity data on β″-alumina single crystals have been meagre, but with recent advances [5.12] in crystal growing more results should be forthcoming. The reasons for the increased conductivity probably include I) the higher Na^+ concentration, II) a lower activation energy because β″-alumina contains two identical sites, and III) a higher mobility because β-alumina probably contains Al^{3+} vacancies and/or O^{2-} interstitials for charge compensation which impede movement of Na^+ ions. The relative importance of each of these factors will not be known until more conductivity data on both β- and β″-alumina are available on samples with controlled compositions and detailed knowledge of the defect concentrations in them.

5.3.4 Conductivity of Polycrystalline β-Alumina

Practical applications will most likely utilize sintered polycrystalline β-alumina, and therefore most conductivity studies have been carried out on polycrystalline samples. Because β-alumina conducts in only two dimensions one might anticipate large grain boundary resistances as Na^+ ions attempt to hop from the conduction plane of one crystallite to the conduction plane of another crystallite. These effects are negligible for isotropic silver ion conductors such as $RbAg_4I_5$ [5.79]. In addition, even if hopping from one crystallite to the next were facile (might be achieved for material sintered to theoretical density) the path length for ion migration from one electrode to the other would be longer than in a properly oriented single crystal. This latter effect, called tortuosity, would decrease the apparent conductivity based on sample geometry, only by a 2/3 factor for a totally randomly oriented sample, but might be somewhat different depending on pressing and sintering conditions. This is especially true when hot-pressing techniques, in which partial orientation takes place, are used [5.18, 71, 72]. The lower conductivity of polycrystalline samples is readily seen in Fig. 5.8 in which single crystal and polycrystalline conductivities are compared for β- and β″-alumina.

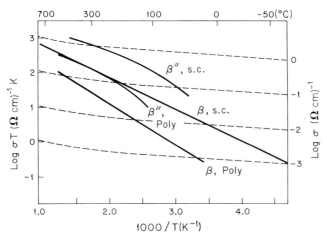

Fig. 5.8. Conductivity of single crystal (s.c.) and polycrystalline (Poly) samples of β- and β″-alumina

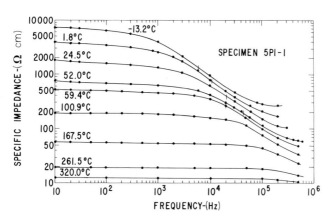

Fig. 5.9. Specific impedance vs. frequency curves for β-alumina specimen at various temperatures [5.81] (Reprinted by permission of the publisher: The Electrochemical Society Inc.)

Unless polycrystalline β-alumina is sintered to nearly theoretical density (~ 3.26 g/cm^3 depending on actual composition) the conductivity is extremely low. Several additives have been proposed, especially in the early literature, to aid the sintering process such as B_2O_3 [5.80], SiO_2 [5.10, 80], ZrO_2 [5.16, 81], Y_2O_3 [5.81–83], and TiO_2 [5.84, 85]. However, it is now recognized that "pure" β-alumina can be sintered to nearly theoretical density if impurities normally present (Si and Ca) are limited to low concentration [5.10]. Sintering of β″-alumina can also be carried out with no additive other than Mg or Li to stabilize the structure [5.86].

To study grain boundary effects it is necessary to separate bulk electrolyte resistance from grain boundary impedance (conceptually equivalent to a parallel

$R - C$ circuit). One approach is to extrapolate 2-electrode AC impedances to infinite frequency to eliminate both electrode/electrolyte and grain boundary effects [5.26, 87] or to use complex plane analysis [5.88]. A more informative method in which 4-electrode AC measurements were taken was developed by *Powers* and *Mitoff* [5.81]. Electrode/electrolyte impedances occurred only at the two outer electrodes so that frequency effects noted at the two inner probes could be attributed to grain boundary impedances. The low frequency limit gave the sum of the bulk and grain boundary resistances while the high frequency limit gave the bulk resistance. Typical data are shown in Fig. 5.9 [5.81]. The high frequency limit at 25 °C was about 84 Ω cm (0.012 $(\Omega \, cm)^{-1}$) while single crystal values have been reported to be 30–72 Ω cm (0.014–0.033 $(\Omega \, cm)^{-1}$). The difference can be attributed to tortuosity and to differences in composition, particularly sodium content. Many investigators have not analyzed their samples after sintering, and sodium content has been shown to differ significantly from the amount of sodium present before sintering [5.26].

Another point established by *Powers* and *Mitoff* [5.81] and discussed by *Bush* [5.89] was the reason for differences in activation energies between single crystals (3.8 kcal/mole) and polycrystalline samples (4–6 kcal/mole) as can be seen in Fig. 5.8. They found the bulk electrolyte resistance activation energy to be 4.4 kcal/mole while the grain boundary activation energy was 6.6 kcal/mole. Thus, one reason for the wide variability in activation energies reported by various workers has been the extent of grain boundary resistance on their measurements.

Figure 5.8 shows that the conductivity of polycrystalline β''-alumina is greater than that of polycrystalline β-alumina by about a 5:1 ratio. At the same time polycrystalline β''-alumina is 5–10 times less conductive than single crystal β''-alumina depending on temperature, because the activation energies are not equal. Because high conductivity is an advantage for practical applications, considerable effort has been expended to achieve the highest possible conductivities with polycrystalline β''-alumina. *Whalen* et al. [5.90] showed that the conductivity increased from 0.10 to 0.20 $(\Omega \, cm)^{-1}$ at 300 °C as sintering time at 1585° increased. The optimum time was 15–20 min.

One technique for improving the conductivity of polycrystalline β-aluminas is to use hot-pressing [5.18, 19, 71, 72, 86, 91]. *Virkar* et al. [5.18] showed that after hot pressing, the β''-alumina contained significant amounts of β-alumina which decreased the conductivity. The β-alumina could be converted to β''-alumina by annealing for 24 h at 1400 °C. During this process the conductivity increased to 0.25 $(\Omega \, cm)^{-1}$ at 300 °C. Probably the highest reported conductivities for hot-pressed β''-alumina are those given by *Weiner* [5.91]. After hot-pressing Li_2O-stabilized β''-alumina at 1440 °C (4260 psi), the conductivity was initially 0.13 $(\Omega \, cm)^{-1}$ at 300 °C. However, after annealing the sample for 20 h at 1450 °C the conductivity reached 0.40 $(\Omega \, cm)^{-1}$ at 300 °C. This point would lie about midway between the single crystal and polycrystalline points on Fig. 5.8.

5.3.5 Conductivity of Spinel Block Doped β-Alumina

Earlier it was shown (Table 5.3) that the conductivity of β-alumina depended on the mobile ion present. It has also been known for many years that metal ions, e.g., Mg^{2+}, which may substitute for Al^{3+} in the spinel block can also affect conductivity [5.16, 73]. Charge balance in undoped β-alumina has been attributed to Al^{3+} vacancies and to O^{2-} interstitials in the conduction plane. If a metal ion having a charge less than $+3$ replaced an aluminum ion, then additional sodium ions could be incorporated for charge balance. Sodium content and conductivity as a function of MgO doping level were studied by *Kennedy* and *Sammells* [5.26]. It was found (Fig. 5.10) that the optimum concentration of sodium (minimum resistivity) depended on MgO content and that as the MgO content increased the optimum sodium content also increased. The optimum sodium content could be predicted by assuming that each Mg^{2+} added incorporated one extra Na^+ compared to the undoped material. Doping with other $+2$ metal ions also increased conductivity [5.73] compared to similar doping with $+3$ metal ions as shown in Table 5.4 [5.92]. However, the effects were smaller than with Mg^{2+}. There are several roles the dopant may play so that the situation may be more complicated than a simple addition of one additional conductive Na^+ for every M^{2+} added, especially when dealing with polycrystalline materials. There is some evidence [5.1] that M^{2+} can exchange with Na^+ in the conduction plane and in this role would be expected to decrease the conductivity. Calcium decreases β-alumina conductivity dramatically with a 1% addition causing a 50–100-fold decrease [5.10, 16, 73]. *De Jonghe* [5.10] attributed this to calcium accumulating in the grain boundary region. Mössbauer [5.93] and diffusion measurements [5.94] have indicated that Eu^{2+} may reside in both the conduction plane and the spinel block. Lithium is another element which has been shown by nmr line narrowing studies [5.40] to reside in

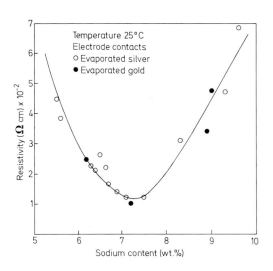

Fig. 5.10. Resistivity as a function of sodium oxide content for β-alumina/2% MgO pellets at 25°C [5.26]

Table 5.4. Conductivity of 2 wt.-% transition metal-doped β-alumina (polycrystalline samples)

Dopant	Oxidation state	Conductivity $[(\Omega\,cm)^{-1}]$
None	—	$5.5\cdot10^{-4}$
Cr	+3	$4.7\cdot10^{-4}$
Fe	+3 (small amt +2)	$6.7\cdot10^{-4}$
Co	+2	$8.0\cdot10^{-4}$
Mn	+2	$10.4\cdot10^{-4}$
Ni	+2	$11.5\cdot10^{-4}$
Mg	+2	$80\ \cdot10^{-4}$

both the conduction plane and the spinel block. When a $+2$ metal ion such as Mg^{2+} substitutes for Al^{3+} it can play at least three roles aside from any grain boundary effects:

1) *Incorporate additional conductive* Na^+

$$NaAl_{11}O_{17}+xMg^{2+}+xNa^+\rightarrow Na_{1+x}Mg_xAl_{11-x}O_{17}+xAl^{3+}$$
defect reaction: $O \rightarrow xNa_i^{\cdot}+xMg'_{Al}$.

(5.6)

2) *Remove interstitial* O^{2-} *which block* Na^+ *conduction*

$$Na_{1+x}Al_{11}O_{17+x/2}+xMgO\rightarrow Na_{1+x}Mg_xAl_{11-x}O_{17}+1/2xAl_2O_3$$
defect reaction: $xNa_i^{\cdot}+x/2O''_i\rightarrow xNa_i^{\cdot}+xMg'_{Al}$.

(5.7)

3) *Remove* Al^{3+} *vacancies which hold* Na^+ *electrostatically* (e.g., $V'''_{Al}-Na_i^{\cdot}$ pairs)

$$Na_{1+x}Al_{11-x/3}O_{17}+xMgO\rightarrow Na_{1+x}Mg_xAl_{11-x}O_{17}+x/3Al_2O_3$$
defect reaction: $xNa_i^{\cdot}+x/3V'''_{Al}\rightarrow xNa_i^{\cdot}+xMg'_{Al}$.

(5.8)

It is possible to produce β-alumina analogs in which all the Al^{3+} ions have been replaced by another M^{3+}. The gallium analogs, $NaGa_{11}O_{17}$ and $KGa_{11}O_{17}$, were prepared by *Foster* and *Stumpf* in 1951 [5.95] and described more recently by *Foster* and *Scardefield* [5.96]. A gallium analog of β"-alumina was reported by *Foster* and *Arbach* [5.97] and *Foxman* et al. [5.98]. Conductivity data were given by *Brinkhoff* [5.99] who studied gallium substitution in β-alumina from $MAl_{11}O_{17}$ to $MGa_{11}O_{17}$ where M was Na^+ or K^+. The results (Fig. 5.11) show that conductivity for the sodium compounds was not affected appreciably by gallium substitution while a marked decrease was observed for the potassium compounds.

The iron analogs of β-alumina are known for potassium, $KFe_{11}O_{17}$ [5.100–102]. The stoichiometry for $K_2O\cdot nFe_2O_3$ ranged from n = 5–11. *Roth* and *Romanczuk* [5.100] showed that potassium ferrite was a mixed conductor with

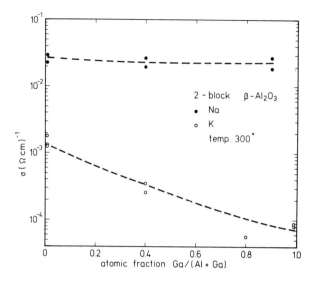

Fig. 5.11. Conductivity as a function of gallium substitution in 2-block β-alumina [5.99]

an activation energy of 6.1 kcal/mole for ionic conductivity and 2.2 kcal/mole for electronic conductivity. The ionic activation energy agreed reasonably well with the potassium diffusion value of 7.2 kcal/mole and also agreed with the activation energy for conduction in K^+-β-alumina (6.8 kcal/mole). *Takahashi* et al. [5.101] also reported a mixed conductivity of 0.5 $(\Omega cm)^{-1}$ at 100 °C for $K_2O \cdot 6–7 Fe_2O_3$. *Dudley* et al. [5.102] were able to change the conductivity at 150 °C by adding K^+ in $K_{1+x}Fe_{11}O_{17}$ using a coulometric titration in the cell:

$$K/K^+\text{-β-alumina}/K_{1+x}Fe_{11}O_{17}.$$

5.3.6 Electronic Conductivity in β-Alumina

The possibility that an additive to β-alumina may lead to electronic conductivity would prevent its use in any battery application using β-alumina as the electrolyte (as part of the anode or cathode structure it could, however, be an advantage). The results for the iron analogs of β-alumina [5.100–102] point out the need for electronic conductivity data on β-alumina and doped β-alumina. Few data exist except for the low levels of electronic conductivity reported by *Whittingham* and *Huggins* [5.76] for Ag^+-β-alumina using a Wagner blocking electrode technique with the cell

$$- Ag/Ag^+\text{-β-alumina}/Pt + .$$

The electronic carrier was considered to be the electron (as opposed to electron hole) and had a conductivity of about $10^{-5} (\Omega cm)^{-1}$ at ~800 °C compared to an ionic conductivity of 0.2 $(\Omega\,\text{cm})^{-1}$. Electronic conductivity was negligible below 500 °C. *Sammells* [5.103] carried out similar measurements with the cell

$$- \text{Na/Na}^+\text{-β-alumina/Fe} +$$

and found the electronic conductivity to be $<3\cdot 10^{-8}$ $(\Omega\,\text{cm})^{-1}$ at 120 °C. The true value will be lower because equilibrium still had not been achieved after 100 h under applied voltage. High doping levels of Fe^{3+} $(\text{NaFe}_{5.5}\text{Al}_{5.5}\text{O}_{17})$ increased electronic conductivity to $6.7 \cdot 10^{-4}$ $(\Omega\,\text{cm})^{-1}$ at 120 °C [5.103]. However, the measurement was taken using two blocking electrodes (Au) so that the electron activity was not fixed at one electrode as it is in the usual Wagner method. *Galli* et al. [5.104] showed the electronic transport number to be essentially zero for undoped, 0.5 wt.% magnesium-doped, 3.8 wt.% Fe_2O_3-doped, and 3.57 wt.% CoO-1.28 wt.% TiO_2-doped β-alumina.

5.3.7 Conduction Mechanism in β-Alumina

Figure 5.5 illustrates the arrangement of mobile ions in the conduction plane of β-alumina which has been deduced from the many structural studies. It should be kept in mind that in β″-alumina BR and aBR sites are identical and represent one set of sites which would be completely filled for stoichiometric $\text{Na}_2\text{MgAl}_{10}\text{O}_{17}$. However, this material is sodium ion deficient, and vacancies in this set of sites will be present. As was pointed out previously, Na^+ distributes between BR and mO positions while Ag^+ distributes between BR and aBR sites in Ag^+-β-alumina.

If BR sites are considered as lattice sites while aBR or mO sites are interstitial sites there are three conduction mechanisms which can be postulated for β-alumina:

1) Vacancy

$$\text{Na}_{BR}^+ + V_{BR} \rightarrow V_{BR} + \text{Na}_{BR}^+. \tag{5.9}$$

2) Interstitial (direct interstitial movement)

$$\text{Na}_{mO}^+ + V_{mO} \rightarrow V_{mO} + \text{Na}_{mO}^+. \tag{5.10}$$

3) Interstitialcy (indirect interstitial movement)

$$\text{Na}_{mO}^{+\,1} + \text{Na}_{BR}^+ + V_{mO} \rightarrow V_{mO} + \text{Na}_{BR}^+ + \text{Na}_{mO}^{+\,1}. \tag{5.11}$$

One experimental method to help decide which of these mechanisms is responsible for conductivity is to compare tracer diffusion coefficients D_t with

---

[1]   Isotopic label to differentiate Na^+ ions initially in mO and BR sites. Note labeled Na^+ moves from mO to BR while charge moves from mO to neighbor mO.

diffusion coefficients calculated from conductivity measurements D_σ using the Nernst-Einstein equation:

$$D_\sigma = \frac{\sigma RT}{cq^2} \tag{5.12}$$

where c is the concentration of mobile ions and q is their charge. The two diffusion coefficients differ because of different jump distances between tracer movement and charge movement and the correlation factor, f. The diffusion coefficients can be given by the following equations

$$D_t = Ga_t^2 v_t f \tag{5.13}$$

$$D_\sigma = Ga_\sigma^2 v_\sigma \tag{5.14}$$

(where G is a geometric factor $= 1/6$ for cubic three-dimensional motion, a is jump distance, and v is jump frequency)

$$D_t/D_\sigma = \left(\frac{a_t}{a_\sigma}\right)^2 \frac{v_t}{v_\sigma} f = H_R \quad \text{(Haven ratio)}. \tag{5.15}$$

The factor f is introduced for tracer jumps because they may not be completely random, e.g., there will be a correlation between successive steps of a particle moving by a vacancy mechanism even if the steps of the vacancies themselves (charge movement) are uncorrelated [5.105–107]. Thus, D_t contains a correlation factor while D_σ does not. However, it should be noted that these relationships have been developed for defect models which contain a small concentration of defects. Complications arising from materials like β-alumina containing a high concentration of defects have been considered by *Sato* and *Kikuchi* [5.108, 109] who introduced a correlation factor for conduction also (see Sect. 2.4.5).

In general, it is usually assumed that $v_\sigma = v_t$ and then a measurement of H_R can be used to help elucidate the conduction mechanism. For the vacancy mechanism $a_t = a_\sigma$ and $f \cong 0.3$–0.8, depending on geometrical configuration [5.105]. Direct interstitial jumps are not correlated ($f = 1$) and $a_t = a_\sigma$ so that $H_R = 1$. On the other hand when the conduction mechanism is indirect, $a_t < a_\sigma$ and $f \neq 1$ and $H_R \neq f$ which has caused considerable confusion in the literature[2].

The Haven ratio for Ag$^+$-β-alumina was found to be 0.61 by *Whittingham* and *Huggins* [5.76]/(0.55 using data given in Table 5.3). Because the conductivity increases with increasing M$^+$ concentration, an interstitial-type mechanism is probable. For the β-alumina structure $(a_t/a_\sigma)^2 = 0.33$ and $f = 1.8$ (found by summation of simple jumps and also by an electrical circuit technique [5.110]) so

---

[2]   *Compaan* and *Haven* defined the expression for H_R as the correlation factor incorporating the jump distance ratio into f. Other authors have used D_σ/D_t in their discussions because it is normally ≥ 1 which is convenient.

that $H_R = 0.60$, in excellent agreement with the experimental value. There are still many uncertainties which means that the close agreement between theory and experiment was probably fortuitous. Consider, for example, that the concentration term in the Nernst-Einstein equation was taken as the stoichiometric value of $5.45 \cdot 10^{-3}$ moles/cm^3, but the actual material contained 15% excess, $6.27 \cdot 10^{-3}$ moles/cm^3, and if only interstitial ions should be considered mobile the correct value would be $(0.82 \cdot 10^{-3} + V_{BR})$ moles/cm^3, where V_{BR} takes into account the ions which should be on lattice sites but have moved to interstitial positions. *Roth* [5.39] found for Ag$^+$-β-alumina containing 25% excess Ag$^+$ that the concentration at aBR sites was $2.7 \cdot 10^{-3}$ moles/cm^3. This would make D_σ higher by a factor of two and throw the whole analysis of mechanism in doubt. In addition, the high concentration of defects would lead to an interaction between defects, and an approach like that taken by *Sato* and *Kikuchi* (path probability method) should be considered [5.108, 109, 111]/(see Sect. 2.4). Another amazing feature is that H_R has also been found to be 0.60 for Na$^+$-β-alumina and Tl$^+$-β-alumina (Table 5.3) which contain mobile ions in different sites and distributions [5.110]. In any case H_R does not appear to be 1.0 so that some sort of indirect interstitial movement is most probable.

For theoretical discussion of β- and β''-aluminas see Section 2.4.

5.4 Applications of β-Alumina

5.4.1 Sodium-Sulfur Battery

In recent years considerable effort has been devoted to the development of new high energy batteries. The two most likely applications would be for electric vehicles and load-leveling storage systems for electric utilities. The specific requirements for these two applications differ in detail but both require high energy density (150–200 Wh/kg), low cost, and long life. Vehicle batteries stress high energy and power density while load leveling batteries stress low cost and long life. Several systems are being developed, and one of these is the sodium-sulfur battery utilizing β-alumina as the electrolyte.

A sodium-sulfur cell was first described by *Kummer* and *Weber* [5.112], and the battery concept is covered by several early patents [5.80, 113–116]. Basically the battery consists of a molten sodium anode and a molten sulfur cathode separated by a β-alumina electrolyte. The β-alumina is often fabricated in the form of a closed-end tube which acts as a container for the sodium anode and is immersed in the molten sulfur as shown in Fig. 5.12 [5.117]. The cathode material is a graphite felt impregnated with sulfur since sulfur itself is a nonconductor. The stainless steel outer container serves as the cathode current collector.

The discharge reaction for the cell is

$$2Na + x S \rightarrow Na_2S_x \tag{5.16}$$

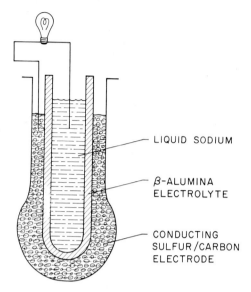

LIQUID SODIUM

β-ALUMINA
ELECTROLYTE

CONDUCTING
SULFUR/CARBON
ELECTRODE

Fig. 5.12. Diagram of sodium-sulfur electrochemical cell [5.117]

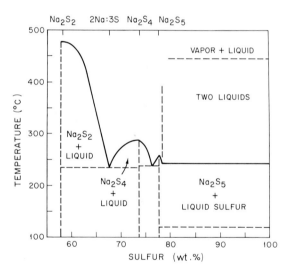

Fig. 5.13. Phase diagram of sodium-sulfur system [5.118]

where x is usually given as 5 when excess sulfur is present but discharge can be carried as far as Na_2S_3 (actually a mixture of Na_2S_4 and Na_2S_2 [5.118]). Beyond this point the polysulfide melting point starts to rise rapidly (Fig. 5.13). Operation of the Na–S battery is usually maintained at ~300 °C, and further discharge would lead to freezing of the cathode material. *Gupta* and *Tischer* [5.119] have investigated the thermodynamic properties and phase diagram for sodium polysulfides and showed that the open-circuit voltage depended on composition and temperature. The value of 2.08 V at 300 °C is normally quoted for systems containing excess sulfur, but the voltage drops to ~1.8 V for compositions approximating Na_2S_3. Theoretical energy storage is 760 Wh/kg which is consistent with the goal of 150–200 Wh/kg in a practical battery since most batteries achieve 20–25% of their theoretical value.

The current status of Na–S battery development is still predominantly in the laboratory although a few large scale tests have been reported. Ford Motor Co. has built and tested a 12 V 24-cell battery containing four parallel strings of six tubular electrolyte cells in series. Peak power of 400 W was achieved resulting in a power density of >209 W/kg and an energy density of >95 Wh/kg excluding thermal insulation and packaging [5.120]. An impressive field test has been carried out by the Electricity Council Research Center. A van was powered by a 960 cell, 50 kWh, 100 V Na–S battery [5.121]. The battery was rated at 15.5 kW average power, 29 kW peak power and weighed 800 kg. Power density was 36 W/kg and energy density was 63 Wh/kg, far short of the 150–200 Wh/kg projected. However, the test proved that the Na–S system can be used for traction power, and assuredly performance will improve as better cells are produced.

The present state of cell design and performance is best summarized by the reports of a National Science Foundation program involving Ford Motor Co., University of Utah, and Rensselaer Polytechnic Institute [5.86, 91, 122]. Cells are normally charged and discharged at 100–200 mA/cm^2 with energy efficiencies of 70–75% at 100 mA/cm^2 and 55–60% at 200 mA/cm^2. One cell has operated 24 months with passage of >1600 Ah/cm^2 in discharge and >9800 cycles. Essentially all performance criteria for the load-leveling operation have been met except the 10–20 year required lifetime, and the next step is a realistic cost analysis to determine whether the Na–S battery can compete with other load leveling schemes. Because of the wider range of operating conditions required, additional laboratory work will be necessary to determine if performance criteria for vehicle traction power can be met.

Concerning life of the β-alumina electrolyte, several studies have shown the existence of various modes of degradation which must be avoided for practical applications. One of the most important questions is whether β″-alumina is superior to β-alumina. There is no doubt that it is more conductive, but several investigators have shown that additives such as magnesium are detrimental to cycle life [5.123, 124]. The aging mechanism involved an ion exchange between impurities (especially K^+) of the reactants and Na^+ in the β-alumina. The preferred Ford Motor Co. material is a β″-alumina stabilized with Li_2O [5.86, 91, 122], although it has been possible to use low cost extrusion methods only for β-alumina with isostatic pressing used for β″-alumina [5.122].

Breakdown of β-alumina electrolyte has also been attributed to formation of molten sodium dendrites during the charging cycle with ultimate electrical shorting [5.125]. *Richman* and *Tennenhouse* [5.126] and *Tennenhouse* et al. [5.127] have studied degradation of β″-alumina during charge and conclude that cracks formed when Na^+ ions migrated through the ceramic and were converted to metallic Na. Their postulated mechanism consisted of ceramic dissolution at preexisting surface cracks which resulted in selective removal of electrolyte from crack tips. The model predicted a threshold current density at which breakdown would occur, and this has been observed experimentally. β''-alumina containing 1.1 wt.-% Li_2O could be charged only at $\leq 200\,mA/cm^2$ while β''-alumina containing 0.25 wt.-% Li_2O could be charged at $\leq 1\,A/cm^2$ before degradation occurred [5.126]. The presence of excess sodium on the fracture surface of β''-alumina from a cell which had failed in operation was detected by Auger electron spectroscopy in support of the proposed failure mechanism that sodium ion migration leads to intergranular weakening and fracture [5.128].

In addition to the references already noted, research on the Na–S battery at various laboratories around the world has been discussed in several review articles [5.129–134].

5.4.2 Other High Temperature Batteries Utilizing β-Alumina

Although the major effort in battery development using β-alumina has been the Na–S battery some research has been directed toward other systems. *Werth* [5.135] patented a secondary battery utilizing a molten alkali metal anode, a porous conductive carbon cathode which adsorbed chlorine, a molten alkali metal chloroaluminate catholyte and a β-alumina electrolyte separator. The discharge reaction for the Na/β-alumina/$NaAlCl_4$, Cl_2, C cell can be written simply as

$$2\,Na + Cl_2 \rightarrow 2\,NaCl\,(in\,NaAlCl_4) \tag{5.17}$$

and could be operated as low as 150–200 °C because of the low melting point of $NaAlCl_4$ with an open-circuit potential of up to 3.5 V depending on the carbon cathode composition.

An extension of this concept was patented by *Werth* [5.136] in which the catholyte contained a reducible metal chloride such as $SbCl_3$, $CuCl_2$, or $FeCl_3$. The discharge reaction would be

$$n\,Na + MCl_x \rightarrow n\,NaCl + MCl_{x-n}. \tag{5.18}$$

The theoretical energy densities were 1034 Wh/kg for $CuCl_2$, 792 Wh/kg for $FeCl_3$, and 825 Wh/kg for $SbCl_3$. *Miller* [5.137] described a similar approach with a Na/β-alumina/$CuCl_2$, $NaAlCl_4$, W cell operating between 136° and 200 °C. However, the cell was not rechargeable, and severe polarization was observed during discharge.

Kennedy and *Sammells* [5.103, 138] and *Kennedy* and *Akridge* [5.92] have investigated cells based on the concept of doping β-alumina with transition metal ions to act as the anode and/or cathode materials. One cell was of the type Na(l)/β-alumina/Fe_2O_3-doped β-alumina and operated in both charge and discharge directions at 120 °C [5.103, 138].

5.4.3 Low Temperature Batteries Utilizing β-Alumina

The major problems to be overcome in order to design ambient temperature batteries are to develop a satisfactory sodium electrode and a satisfactory cathode material. The solid sodium/β-alumina interface has presented problems [5.103, 138], and therefore a liquid alloy of sodium or an organic liquid electrolyte between sodium and β-alumina has usually been employed. Cathode materials which have been investigated include aqueous solutions of halogens, metal halides, oxygen, or water itself as the electron accepting material. The role of β-alumina is to prevent contact of sodium metal with water and to allow only passage of Na^+ ions from the anode to the cathode. The batteries have usually been pursued only as primary cells. Most of the results have appeared in patents, although a paper by *Will* and *Mitoff* [5.139] summarized the properties of several systems as shown in Table 5.5. Additional systems and references are listed in Table 5.6. The β-alumina/propylene carbonate interface present in some of these cells was studied by *Farrington* [5.155] and *Voinov* and *Tannenberger* [5.156] who found that Na^+ transport may have involved an adsorbed intermediate state and was influenced by the presence of H_2O or H_3O^+ in the β-alumina. The effect of water on β-alumina was discussed in an earlier section.

5.4.4 Miscellaneous Applications of β-Alumina

As early as 1968, *Hever* [5.157] described the use of β-alumina doped with iron and titanium as an electrode material in cells

Fe, Ti-doped β-alumina/β-alumina/Fe, Ti-doped β-alumina .

Passage of current through the cell resulted in an increase in voltage, i.e., the cell acted as a capacitor. *Hever* envisaged its use as an electronic integrator or as a capacitor in filter circuits. *Hever* and *Kummer* also conceived of using two different doped β-aluminas, one doped with an electron donor, the other with an electron acceptor to act as anolyte and catholyte in a secondary battery [5.157, 158]. This concept has been pursued in more detail by *Kennedy* and *Sammells* [5.103, 138], *Kennedy* and *Akridge* [5.92], and *Dudley* et al. [5.102].

Because β-alumina may exhibit a Na^+ transference number of one, it can be used in a specific ion electrode. *Joglekar* et al. [5.159] described a β-alumina probe for sodium metal activity in molten metals. A β-alumina pellet was sealed in an α-alumina tube with Kovar 7052 glass and filled with pure sodium which acted as a reference electrode. The probe was immersed in Na–Sn alloy baths at 500 °C, and the potential was directly proportional to log a_{Na}.

Lundsgaard and *Brook* [5.160] attempted to use β-alumina as a sodium specific ion electrode in aqueous solutions but found that the potentials depended on the presence of other M^+ and H^+ ions which could enter the β-

Table 5.5. Summary of cell performance characteristics of low temperature batteries using β-alumina[a]

Cell couple	Open-circuit voltage [V]	Voltage at 0.1 mA/cm^2 [V]	Energy density	
			W-h/kg theoretical	W-h/kg practical
Na-Hg/Br$_2$, H$_2$O	3.77	3.5	1026	330–440
Na-Hg/Br$_2$, propylene carbonate	3.34	3.1	910	330–440
Na-Hg/I$_2$, H$_2$O	3.25	3.0	579	220–330
Na-Hg/I$_2$, propylene carbonate	2.82	2.8	502	170–260
Na-Hg/O$_2$, NaHCO$_3$	2.90	2.8	1553	660
Na-Hg/H$_2$O, NaOH	1.85	1.7	991	440

[a]   Data taken from Ref. [5.139].

Table 5.6. Ambient temperature primary batteries with β-alumina electrolyte

Cell	Investigators	Ref.
Na(Hg)/β/Br$_2$ or I$_2$ in nitrobenzene	*Will, Farrington*	[5.140]
Na(Hg)/β/Br$_2$ or I$_2$ in propylene carbonate	*Will, Dubin*	[5.141]
Na(Hg)/β/NaOH or NaCl in H$_2$O	*Mitoff, Will*	[5.142]
Na(Hg)/β/Br$_2$ or I$_2$ in POCl$_3$ or POBr$_3$	*Farrington, Will, Lord*	[5.143]
Na(Hg)/β/ICl in acetonitrile, butyrolactone, or propylene carbonate	*Farrington*	[5.144]
Na(Hg)/β/Br$_2$ in tri- or tetraalkylammonium chloride or bromide	*Farrington*	[5.145]
Na(Hg)/β/ICl in SOCl$_2$, POCl$_3$ or SO$_2$Cl$_2$	*Lord, Farrington*	[5.146, 147]
Na(Hg)/β/AlCl$_3$ in SO$_2$Cl$_2$	*Farrington, Lord*	[5.148]
Na(s)/β/NiF$_2$, DMF	*Tannenberger, Charhonnier*	[5.149]
Na/NaClO$_4$ in propylene carbonate/β/AgF, NaClO$_4$ in H$_2$O	*Terazaki*	[5.150]
Na/NaClO$_4$ in THF/β/I$_2$ in H$_2$O	*Terazaki*	[5.151]
Na(s)/β/NiCl$_2$, CuCl$_2$, FeCl$_2$, CuBr$_2$, NiBr$_2$ or NiI$_2$ in H$_2$O	*Kashihara, Terazaki*	[5.152]
Na(s)/β/NiF$_2$, NaClO$_4$ in H$_2$O	*Terazaki*	[5.153]
Na(Hg)/β/Br$_2$, NaBr in propylene carbonate	*Touzain, Bonnetain*	[5.154]

alumina conduction plane. *Lundsgaard* and *Brook* [5.161] were more successful in devising a β-alumina cell sensitive to H$_2$ or O$_2$. The cell was

$$\text{H}_2 \text{ or O}_2 \text{(std), Pt/β-   and   β''-alumina/Pt, H}_2 \text{ or O}_2 \text{(unk)}$$

in which the gas streams were saturated with H$_2$O vapor. However, the response time was slow at 23 °C which made practical application doubtful. On the other hand, *Choudhury* [5.162] was successful in making thermodynamic measurements with β-alumina at 500–1000 °C. Oxygen partial pressures were fixed using metal/metal oxide couples in the cells

Fe, FeO/α-   and   β-alumina/Ni, NiO

Ni, NiO/α-   and   β-alumina/Cu, Cu$_2$O .

Results agreed well with the same measurements taken with an oxide ion conductor as the electrolyte. Similar cells used to determine Na$_2$O activity and aluminum chemical potential were

Ni, NiF$_2$, NaF/α-   and   β-alumina/Cu, Cu$_2$O

Co, CoF$_2$, NaF/α-   and   β-alumina/Ni, NiF$_2$, NaF

Al/α-   and   β-alumina/AlF$_3$, NiF$_2$, Ni

Al/α-   and   β-alumina/NaF, NiF$_2$, Ni .

Choudhury concluded that β-alumina could be used to measure extremely low oxygen partial pressure at lower temperatures than oxide ion conducting electrolytes. A patent has been issued for the use of β-alumina in an oxygen sensor [5.163]. The oxygen sensitive electrode consisted of a sintered Ag gas diffusion cathode coated with a gas-permeable poly(tetrafluoroethylene) film. A β-alumina tube was filled with either a zinc or sodium anode and current was related to oxygen pressure in the gas stream. Thermodynamic measurements have also been made on Ag$_5$Te$_3$ using Ag$^+$-β-alumina by *Kutsenok* et al. [5.164] who determined ΔH_f^0 and S_f^0 (250–350 °C) with the cell Ag/Ag$^+$-β-alumina/Te, Ag$_5$Te$_3$/Pt.

The high transference number for Na$^+$ was used by *Kummer* and *Weber* [5.165] in devising an electrochemical separation of Na from a mixture of alkali metal salts. It would seem from other studies that because other alkali metal ions can enter, the β-alumina would limit the use of this device.

A thermoelectric device based on β-alumina has been described by *Weber* [5.166] and possesses several attractive features. Under ideal no load conditions the thermodynamic efficiency approaches the Carnot limit and electrical power outputs of 0.5 W/cm^2 appear achievable. If an electromagnetic pump or wick were used for sodium metal, the only moving part would be the circulating sodium. Practical utility will depend on the long term durability of β-alumina in liquid sodium or sodium vapor which is unknown at this time.

References

5.1   Y.F.Y.Yao, J.T.Kummer: J. Inorg. Nucl. Chem. **29**, 2453 (1967)
5.2   G.A.Rankin, H.E.Merwin: J. Am. Chem. Soc. **38**, 568 (1916)
5.3   R.Ridgway, A.Klein, W.O'Leary: Trans. Electrochem. Soc. **70**, 71 (1936)

5.4   S.B.Hendricks, L.Pauling: Z. Krist. **64**, 303 (1927)
5.5   W.L.Bragg, C.Gottfried, J.West: Z. Krist. **77**, 255 (1931)
5.6   C.A.Beevers, S.Brohult: Z. Krist. **95**, 472 (1936)
5.7   C.A.Beevers, M.A.S.Ross: Z. Krist. **97**, 59 (1937)
5.8   J.T.Kummer: Progr. Solid State Chem. **7**, 141 (1972)
5.9   N.Weber, A.Venero: Am. Ceram. Soc. Meeting, Paper 4-J1-70, May 1970. See Ref. 5.8
5.10   L.C.DeJonghe: EPRI Rept. 252, July 1975
5.11   A.K.Ray, E.C.Subbarao: Mat. Res. Bull. **10**, 583 (1975)
5.12   A.D.Morrison, R.W.Stormont, F.H.Cocks: J. Am. Ceram. Soc. **58**, 41 (1975)
5.13   F.H.Cocks, R.W.Stormont: J. Electrochem. Soc. **121**, 596 (1974)
5.14   R.J.Baughman, R.A.Lefever: Mat. Res. Bull. **10**, 607 (1975)
5.15   P.J.Yancey: U.S. Patent 3917462, Nov. 4, 1975
5.16   I.Wynn Jones, L.J.Miles: Proc. Brit. Ceram. Soc. **19**, 161 (1971)
5.17   R.L.Clendenen, E.E.Olson: U.S. Patent 3795723, March 5, 1974
5.18   A.V.Virkar, G.J.Tennenhouse, R.S.Gordon: J. Am. Ceram. Soc. **57**, 508 (1974)
5.19   W.J.McDonough, R.W.Rice: "Hot Pressing and Physical Properties of Sodium Beta-Alumina"; Rept. NRL Prog., Naval Res. Lab. (1975)
5.20a Y.Lazennec, J.Fally: Fr. Demande 2092845, March 3, 1972
5.20b J.Fally, C.Lasne, Y.Lazennec, Y.le Cars, P.Margotin: J. Electrochem. Soc. **120**, 1296 (1973)
5.21   R.W.Powers: J. Electrochem. Soc. **122**, 490 (1975)
5.22a R.W.Powers, R.S.Owens: U.S. Patent 3896018, July 22, 1975
5.22b R.W.Powers, R.A.Giddings: U.S. Patent 3881661, May 6, 1975
5.23   J.H.Kennedy, A.Foissy: J. Electrochem. Soc. **122**, 482 (1975)
5.24   J.H.Kennedy, A.Foissy: J. Am. Ceram. Soc. **60**, 33 (1977)
5.25   T.L.Markin, R.J.Bones, J.L.Woodhead, L.L.Wassell: Brit. Patent 1386244, March 5, 1975
5.26   J.H.Kennedy, A.F.Sammells: J. Electrochem. Soc. **119**, 1609 (1972)
5.27   T.Sasamoto, T.Sata: Yogyo Kyokai Shi **82**, 420 (1974)
5.28   H.Saalfeld, H.Matthies, S.K.Datta: Ber. Deut. Keram. Ges. **45**, 212 (1968)
5.29   D.Kline, H.S.Story, W.L.Roth: J. Chem. Phys. **57**, 5180 (1972)
5.30   F.G.Will: J. Electrochem. Soc. **123**, 834 (1976)
5.31   G.C.Farrington: J. Electrochem. Soc. **123**, 833 (1976)
5.32   A.Yamaguchi, E.Kato: Yogyo Kyokai Shi **83**, 485 (1975)
5.33   G.Yamaguchi: J. Japan Electrochem. Soc. **11**, 260 (1943)
5.34   G.Yamaguchi, K.Suzuki: Bull. Chem. Soc. Japan **41**, 93 (1968)
5.35   M.Bettman, C.R.Peters: J. Phys. Chem. **73**, 1774 (1969)
5.36   M.Bettman, L.L.Terner: Inorg. Chem. **10**, 1442 (1971)
5.37   J.Felsche: Naturwissenschaften **54**, 612 (1967)
5.38   C.R.Peters, H.Bettman, J.W.Moore, M.D.Glick: Acta Cryst. B **27**, 1826 (1971)
5.39   W.L.Roth: J. Solid State Chem. **4**, 60 (1972)
5.40   W.L.Roth: Trans. Am. Cryst. Assoc. **11**, 51 (1975)
5.41   P.D.Dernier, J.P.Remeika: J. Solid State Chem. **17**, 245 (1976)
5.42   T.Kodama, G.Muto: J. Solid State Chem. **17**, 61 (1976)
5.43   W.L.Roth, W.C.Hamilton, S.J.La Placa: Am. Cryst. Assoc. Abstr., Ser. 2, **1**, 169 (1973)
5.44   W.L.Roth, F.Reidinger, S.J.La Placa: Studies of Stabilization and Transport Mechanisms in Beta and Beta″-Alumina by Neutron Diffraction. In *Superionic Conductors*, ed. by G.D.Mahan, W.L.Roth (Plenum Press, New York, London 1976) pp. 223–241
5.45   J.P.Boilot, G.Collin, R.Comès, J.Théry, R.Collongues, A.Guinier: x-Ray Diffuse Scattering from β-Alumina. In *Superionic Conductors*, ed. by G.D.Mahan, W.L.Roth (Plenum Press, New York, London 1976) pp. 243–260
5.46   Y. Le Cars, R.Comès, L.Deschamps, J.Théry: Acta Cryst., Sect. A, **30**, 305 (1974)
5.47   D.B.McWhan, P.D.Dernier, C.Vettier, J.P.Remeika: x-Ray Diffuse Scattering from Alkali and Europium β-Aluminas. In *Superionic Conductors*, ed. by G.D.Mahan, W.L.Roth (Plenum Press, New York, London 1976) p. 376

5.48 D.B.McWhan, S.M.Shapiro, J.P.Remeika, G.Shiranes: J. Phys. C. Solid State Phys. **8**, L487 (1975)

5.49 S.M.Shapiro: Neutron Scattering Studies of Solid Electrolytes. In *Superionic Conductors*, ed. by G.D.Mahan, W.L.Roth (Plenum Press, New York, London 1976) pp. 261–277

5.50 B.C.Tofield, A.J.Jacobson, W.A.England: Powder Neutron Diffraction Analysis of Sodium Silver and Deuterium Beta Aluminas. In *Superionic Conductors*, ed. by G.D.Mahan, W.L.Roth (Plenum Press, New York, London 1976) pp. 375–376

5.51 D.Jérome, J.P.Boilot: J. Phys. (Paris) Letters **35**, L129 (1974)

5.52 I.Chung, H.S.Story, W.L.Roth: Bull. Am. Phys. Soc. Series II **19**, 202 (1974)

5.53 I.Chung, H.S.Story, W.L.Roth: J. Chem. Phys. **63**, 4903 (1975)

5.54 W.Bailey, S.Glowinkowski, H.Story, W.L.Roth: J. Chem. Phys. **64**, 4126 (1976)

5.55 R.E.Walstedt, R.Dupree, J.P.Remeika: Nuclear Relaxation and Barrier Height Distribution in Na β-Alumina. In *Superionic Conductors*, ed. by G.D.Mahan, W.L.Roth (Plenum Press, New York, London 1976) p. 369

5.56 R.D.Armstrong, P.M.A.Sherwood, R.A.Wiggins: Spectrochim. Acta **30**A, 1213 (1974)

5.57 W.M.Risen,Jr., W.M.Butler: Far Infrared Spectra of Sodium-β-Alumina and Its K^+, Rb^+, Cs^+, Ag^+, and Tl^+ Analogs. In *Superionic Conductors*, ed. by G.D.Mahan, W.L.Roth (Plenum Press, New York, London 1976) pp. 372–373

5.58 S.J.Allen,Jr., J.P.Remeika: Phys. Rev. Letters **33**, 1478 (1974)

5.59 L.L.Chase, C.H.Hao, G.D.Mahan: Solid State Commun. **18**, 401 (1976)

5.60 W.L.Roth: NBS Special Publication **364**, 129 (1972)

5.61 D.J.M.Bevan, B.Hudson, P.T.Moseley: Mat. Res. Bull. **9**, 1073 (1974)

5.62a L.C.De Jonghe: J. Mater. Sci. **10**, 2173 (1975)

5.62b L.C.De Jonghe: J. Mater. Sci. **11**, 206 (1976)

5.63 P.Rao: Electrochem. Soc. 149th Meeting Extended Abstracts, Washington, D.C., 215 (1976)

5.64 N.A.Toropov, M.M.Stukalova: C. R. Akad. Sci. SSSR **24**, 459 (1939)

5.65 N.A.Toropov, M.M.Stukalova: C. R. Akad. Sci. SSSR **27**, 974 (1940)

5.66 R.H.Radzilowski, J.T.Kummer: Inorg. Chem. **8**, 2531 (1969)

5.67 R.H.Radzilowski: Inorg. Chem. **8**, 994 (1969)

5.68 M.S.Whittingham, R.W.Helliwell, R.A.Huggins: U.S. Govt. Res. Dev. Rept. **69**, 228 (1969)

5.69 T.Nishikawa, T.Nishida, Y.Katsuyama, Y.Kawakita, I.Vei: Nippon Kagaku Kaishi 1048 (1974)

5.70 R.H.Radzilowski, Y.F.Yao, J.T.Kummer: J. Appl. Phys. **40**, 4716 (1969)

5.71 B.J.Dunbar, S.Sarian: American Ceramic Soc. 28th Pacific Coast Regional and Nuclear Div. Meeting, 4-FC-75P, October 29–31, 1975

5.72 R.Terai, I.Ogino, H.Wakabayashi: Osaka Kogyo Gijutsu Shikensho Kiho **26**, 52 (1975)

5.73a A.Imai, M.Harata: Electrochem. Soc. Meeting, Abst. 277, May 1970

5.73b A.Imai, M.Harata: Japan. J. Appl. Phys. **11**, 180 (1972)

5.74 M.S.Whittingham, R.A.Huggins: J. Chem. Phys. **54**, 414 (1971)

5.75 W.L.Fielder, H.E.Kautz, J.S.Fordyce, J.Singer: J. Electrochem. Soc. **122**, 528 (1975)

5.76a M.S.Whittingham, R.A.Huggins: NBS Special Publication **364**, 139 (1972)

5.76b M.S.Whittingham, R.A.Huggins: J. Electrochem. Soc. **118**, 1 (1971)

5.77 R.H.Radzilowski, J.T.Kummer: J. Electrochem. Soc. **118**, 714 (1971)

5.78 K.Itoh, K.Kondo, A.Sawaoka, S.Saito: Japan. J. Appl. Phys. **14**, 1237 (1975)

5.79 D.O.Raleigh: J. Appl. Phys. **41**, 1876 (1970)

5.80 J.T.Kummer, N.Weber: U.S. Patent 3404036, October 1, 1968

5.81 R.W.Powers, S.P.Mitoff: J. Electrochem. Soc. **122**, 226 (1975)

5.82 S.Imai, R.Watanabe, Y.Ogawa: Japan Patent 73-31731, October 1, 1973

5.83 R.J.Charles, S.P.Mitoff, W.G.Morris: U.S. Patent 3607435, September 21, 1971

5.84 K.Nishimura, M.Hasegawa, M.Takagi: Japan Patent 73-43646, December 20, 1973

5.85 I.Kovachev: Teor. Tekhnol. Spekaniya, Dokl., Mezhdunar. Kollok., 2nd, 249 (1974)

5.86 Ford Motor Co.: Annual Rept., NSF Contract NSF-C805, July, 1974

5.87 R.D.Armstrong, T.Dickinson, J.Turner: J. Electrochem. Soc. **118**, 1135 (1971)

5.88   R. D. Armstrong, T. Dickinson, J. Turner: Electroanal. Chem. and Interfac. Electrochem. **44**, 157 (1973)

5.89   J. B. Bush, Jr.: EPRI Interim Rept. 128-2, Sept. 1975

5.90   T. J. Whalen, G. J. Tennenhouse, C. Meyer: J. Am. Ceram. Soc. **57**, 497 (1974)

5.91   S. A. Weiner: RANN Rept., Contract NSF-C805 (AER-73-07199), July 1975

5.92   J. H. Kennedy, J. R. Akridge: Electrochem. Soc. Meeting, Washington, D.C., Extended Abstracts p. 223 (1976)

5.93   R. L. Cohen, J. P. Remeika, K. W. West: J. Phys. (Paris), Colloq. **6**, 513 (1974)

5.94   D. B. McWhan, S. J. Allen, Jr., J. P. Remeika, P. D. Dernier: Phys. Rev. Letters **35**, 953 (1975)

5.95   L. M. Foster, H. C. Stumpf: J. Am. Chem. Soc. **73**, 1591 (1951)

5.96   L. M. Foster, J. E. Scardefield: J. Electrochem. Soc. **123**, 141 (1976)

5.97   L. M. Foster, C. V. Arbach: Reaction of β''-Sodium Gallate with Water. In *Superionic Conductors*, ed. by G. D. Mahan, W. L. Roth (Plenum Press, New York, London 1976) pp. 420–421

5.98   B. M. Foxman, S. J. La Placa, L. M. Foster: Mechanism of Formation of the Sodium Gallate Superionic Conductors. In *Superionic Conductors*, ed. by G. D. Mahan, W. L. Roth (Plenum Press, New York, London 1976) pp. 419–420

5.99   H. C. Brinkhoff: J. Phys. Chem. Solids **35**, 1225 (1974)

5.100 W. L. Roth, R. J. Romanczuk: J. Electrochem. Soc. **116**, 975 (1969)

5.101 T. Takahashi, K. Kuwabara, Y. Kase: Denki Kagaku **43**, 273 (1975)

5.102 G. J. Dudley, B. C. H. Steele, A. T. Howe: J. Solid State Chem. **18**, 141 (1976)

5.103a A. F. Sammells: Ph.D. Thesis, University of California, Santa Barbara, 1973, p. 101

5.103b J. H. Kennedy, A. F. Sammells: J. Electrochem. Soc. **121**, 1 (1974)

5.104 R. Galli, P. Longhi, T. Massini, F. A. Tropeano: Electrochim. Acta **18**, 1013 (1973)

5.105 K. Compaan, Y. Haven: Trans. Faraday Soc. **52**, 786 (1956)

5.106 K. Compaan, Y. Haven: Trans. Faraday Soc. **54**, 1498 (1958)

5.107 A. D. LeClaire: Random Walk Theory of Diffusion and Correlation Effects. In *Fast Ion Transport in Solids*, ed. by W. van Gool (North Holland/American Elsevier, Amsterdam 1973) pp. 51–79

5.108 H. Sato, R. Kikuchi: J. Chem. Phys. **55**, 677 (1971)

5.109 R. Kikuchi, H. Sato: J. Chem. Phys. **55**, 702 (1971)

5.110 M. S. Whittingham: Electrochim. Acta **20**, 575 (1975)

5.111 H. Sato, R. Kikuchi: Path Probability Method as Applied to Problems of Superionic Conduction. In *Superionic Conductors*, ed. by G. D. Mahan, W. L. Roth (Plenum Press, New York, London 1976) pp. 135–142

5.112 J. T. Kummer, N. Weber: Proc. Power Sources Conf. **21**, 37 (1967)

5.113 J. T. Kummer, N. Weber: U.S. Patent 3 413 150, Nov. 26, 1968

5.114 J. T. Kummer: U.S. Patent 3 468 709, Sept. 23, 1969

5.115 G. J. Tennenhouse: U.S. Patent 3 446 677, May 27, 1969

5.116 R. W. Minck: U.S. Patent 3 514 332, May 26, 1970

5.117 T. J. Whalen: SAE (Tech. Pap.) 750373 (1975)

5.118 D. G. Oei: Inorg. Chem. **12**, 435 (1973)

5.119 N. K. Gupta, R. P. Tischer: J. Electrochem. Soc. **119**, 1033 (1972)

5.120 S. Gratch, J. V. Petrocelli, R. P. Tischer, R. W. Minck, T. J. Whalen: Intersoc. Energy Conv. Eng. Conf., Conf. Proc. **7**, 38 (1972)

5.121 L. S. Marcoux, E. T. Seo: Adv. Chem. Series No. 140, American Chem. Soc. (1975) pp. 216–227

5.122 S. A. Weiner, R. P. Tischer: RANN Rept., Contract NSF-C805 (AER-73-07199), July 1976

5.123 J. Fally, C. Lasne, Y. Lazennec, P. Margotin: J. Electrochem. Soc. **120**, 1292 (1973)

5.124 Y. Lazennec, C. Lasne, P. Margotin, J. Fally: J. Electrochem. Soc. **122**, 734 (1975)

5.125 R. D. Armstrong, T. Dickinson, J. Turner: Electrochim. Acta **19**, 187 (1974)

5.126 R. H. Richman, G. J. Tennenhouse: J. Am. Ceram. Soc. **58**, 63 (1975)

5.127 G. J. Tennenhouse, R. C. Ku, R. H. Richman, T. J. Whalen: Am. Ceram. Soc. Bull. **54**, 523 (1975)

5.128 C. T. H. Stoddart, E. D. Hondros: Trans. and J. Brit. Ceram. Soc. **73**, 61 (1974)

5.129 M. D. Hames, J. H. Duncan: SAE (Tech. Pap.) 750375 (1975)

5.130 S.P.Mitoff, J.B.Bush,Jr.: Intersoc. Energy Convers. Eng. Conf. Proc. **9**, 916 (1974)

5.131 J.L.Sudworth: IEEE 10th Conf. (1975) pp. 616–620

5.132 N.Weber: Applications of Beta Alumina in the Energy Field. In *Superionic Conductors*, ed. by G.D.Mahan, W.L.Roth (Plenum Press, New York, London 1976) pp. 37–46

5.133 R.Galli, F.Garbassi: Chim. Ind. (Milan) **55**, 522 (1973)

5.134 J.R.Birk: Energy Storage, Batteries, and Solid Electrolytes: Prospects and Problems. In *Superionic Conductors*, ed. by G.D.Mahan, W.L.Roth (Plenum Press, New York, London 1976) pp. 1–14

5.135 J.J.Werth: U.S. Patent 3847667, Nov. 12, 1974

5.136 J.J.Werth: U.S. Patent 3877984, Apr. 15, 1975

5.137 R.O.Miller: NASA Tech. Memo. TMS-3245 (1975)

5.138 J.H.Kennedy, A.F.Sammells: Preparation of Highly Conductive β-Alumina and Electrochemical Measurements. In *Fast Ion Transport in Solids*, ed. by W.van Gool (North Holland/American Elsevier, Amsterdam 1973) pp. 563–572

5.139 F.G.Will, S.P.Mitoff: J. Electrochem. Soc. **122**, 457 (1975)

5.140 F.G.Will, G.C.Farrington: U.S. Patent 3879220, Apr. 22, 1975

5.141 F.G.Will, R.R.Dubin: U.S. Patent 3868273, Feb. 25, 1975

5.142 S.P.Mitoff, F.G.Will: U.S. Patent 3703415, Nov. 21, 1972

5.143 G.C.Farrington, F.G.Will, P.C.Lord: U.S. Patent 3899352, Aug. 12, 1975

5.144 G.C.Farrington: U.S. Patent 3879221, Apr. 22, 1975

5.145 G.C.Farrington: U.S. Patent 3879219, Apr. 22, 1975

5.146 P.C.Lord, G.C.Farrington: U.S. Patent 3879222, Apr. 22, 1975

5.147 G.C.Farrington, P.C.Lord: U.S. Patent 3879223, Apr. 22, 1975

5.148 G.C.Farrington, P.C.Lord: U.S. Patent 3879224, Apr. 22, 1975

5.149 H.Tannenberger, J.Charhonnier: U.S. Patent 3730771, May 1, 1973

5.150 M.Terazaki: Japan. Patent 75131034, Oct. 16, 1975

5.151 M.Terazaki: Japan. Patent 75121744, Sept. 23, 1975

5.152 S.Kashihara, M.Terazaki: Japan. Patent 75118226, Sept. 16, 1975

5.153 M.Terazaki: Japan. Patent 75136627, Oct. 30, 1975

5.154 Ph.Touzain, L.Bonnetain: Electrochim. Acta **19**, 591 (1974)

5.155 G.C.Farrington: J. Electrochem. Soc. **123**, 591 (1976)

5.156 M.Voinov, H.Tannenberger: Electrochim. Acta **19**, 959 (1974)

5.157 K.O.Hever: J. Electrochem. Soc. **115**, 830 (1968)

5.158 K.O.Hever, J.T.Kummer: U.S. Patent 3499796, Mar. 10, 1970

5.159 B.V.Joglekar, P.S.Nicholson, W.W.Smeltzer: Can. Met. Quart. **12**, 155 (1973)

5.160 J.S.Lundsgaard, R.J.Brook: J. Mat. Sci. **8**, 1519 (1973)

5.161 J.S.Lundsgaard, R.J.Brook: J. Mat. Sci. **9**, 2061 (1974)

5.162 N.S.Choudhury: J. Electrochem. Soc. **120**, 1663 (1973)

5.163 United Kingdom Secretary of State for Defense: Ger. Patent 2358491, June 6, 1974

5.164 B.B.Kutsenok, A.R.Kaul, Yu.D.Tret'yakov: Zh. Fiz. Khim. **48**, 2128 (1974)

5.165 J.T.Kummer, N.Weber: Swed. Patent 346223, July 3, 1972

5.166 N.Weber: Energy Conv. **14**, 1 (1974)

6. Oxide Solid Electrolytes

W. L. Worrell

With 13 Figures

Oxide solid electrolytes are the most commonly used solid electrolytes at elevated temperatures. Useful oxide electrolytes are solid solutions which have the fluorite crystal structure and abnormally high oxygen-ion conductivities. Relevant structural characteristics and phase relationships in these solid solutions are summarized in Section 6.1. Defect interactions and conduction mechanisms in a fluorite-type oxide, i.e., ThO_2, are described in Sections 6.2. Various aspects of ionic-conduction processes in oxide solid electrolytes are discussed in Section 6.3. These include conduction mechanisms in $ThO_2(Y_2O_3)$ and $ZrO_2(CaO)$ solid solutions, and effects of composition, temperature and oxygen pressure on the ionic conductivity. The electrochemical principles and techniques necessary for successful use of these electrolytes in high temperature research are described in Section 6.4.1. Applications of oxide electrolytes in oxygen probes and in high-temperature fuel cells are briefly summarized in the last two subsections.

In 1899, *Nernst* [6.1] observed ionic conductivity in a $ZrO_2(Y_2O_3)$ solid solution, and a fuel cell using this oxide electrolyte was first constructed in 1937 [6.2]. However, it was the publication of two papers by *Kiukkola* and *Wagner* in 1957 [6.3, 4] which stimulated intense interest in the oxide electrolytes. Over 600 research articles concerning the properties and applications of these electrolytes were published between 1957 and 1969. The oxide electrolytes have been used not only in high-temperature research [6.5–8] but also in electrochemical probes [6.9–17] and high-temperature fuel cells [6.5, 18–21].

6.1 Structural Characteristics and Phase Relationships

Although other oxides such as BeO, and Y_2O_3 are ionic conductors [6.5, 22], solid solutions formed by Group IV B oxides (ZrO_2, HfO_2) and ThO_2 are the most commonly used oxide electrolytes. The Ti^{+4} oxidation state is not unique in the titanium-oxygen system, and there are a number of stable titanium oxides [6.23]. Thus it is not surprising that TiO_2 exhibits electronic conductivity [6.22, 24] and does not form ionic solid solutions.

Useful oxide electrolytes have the fluorite crystal structure which is shown in Fig. 6.1. The fluorite structure is a face-center-cubic arrangement of cations with the anions occupying all the tetrahedral sites. In this structure each metal cation is surrounded by eight oxygen anions, and each oxygen anion is tetrahedrally

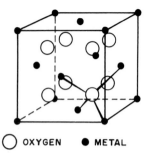

○ OXYGEN   ● METAL   Fig. 6.1. The fluorite crystal structure

coordinated with four metal cations. The fluorite structure has a large number of octahedral interstitial voids. Thus this structure is a rather "open" one, and rapid ion diffusion might be expected. Thoria and a high temperature modification of ZrO_2 have the fluorite crystal structure [6.22]. For ThO_2, *Möebius* [6.25] has calculated that the space available in the fluorite lattice for movement of $O^=$ ions is larger than that available for Th^{+4} ions and has concluded that oxygen ions should diffuse faster than the cations.

However, it is solid solutions such as $ThO_2(Y_2O_3)$ and $ZrO_2(CaO)$ which exhibit abnormally high ionic conductivity. A critical question is how much metal oxide can one dissolve into ThO_2 before the fluorite crystal structure becomes unstable. Happily there is a fairly wide compositional range, over which the fluorite structure is stable in thoria-based solid solutions. For example, the maximum solid solubility limits of CaO and Y_2O_3 in the fluorite phase of ThO_2 are about 10 and 20 mol-%, respectively, at 1700 °C (Table 6.1). At lower temperatures, there are also reasonable solubility limits, although they are smaller and less accurately known.

With the zirconia-based and hafnia-based solid solutions, there is an additional consideration. The fluorite structure is not stable in HfO_2 and is stable in ZrO_2 only above 2300 °C [6.22, 32]. Thus a minimum amount of a metal oxide such as CaO or Y_2O_3 must be present to stabilize the fluorite phase in zirconia and hafnia solid solutions.

Compositional boundaries for the fluorite single-phase in zirconia-calcia and zirconia-yttria solid solutions at 1500 °C are tabulated in Table 6.1. There is controversy concerning the temperature below which the fluorite phase in $ZrO_2(CaO)$ solid solutions becomes unstable. Values for the fluorite decomposition temperature range from 1230 °C [6.28] to 800 °C [6.33]. Extended heat treatments at various temperatures (for example, 2000 h at 815 °C) indicate that the fluorite phase in $ZrO_2(CaO)$ solid solutions is thermodynamically stable above ~900 °C and kinetically stable below this temperature [6.34]. The fluorite phase in $ZrO_2(Y_2O_3)$ solid solutions does not decompose during heat treatments for extended times at temperatures between 800 and 1500 °C [6.29]. In the $ZrO_2(MgO)$ system, the fluorite phase is stable only between 11 and 15 mol-% MgO at 1500 °C and decomposes into tetragonal ZrO_2 and MgO at temperatures below 1400 °C [6.35].

Table 6.1. Fluorite phase boundaries in oxide solid solutions

Oxide Solid Solution	Boundaries[a, b]	Temperature [°C]	Reference
ThO_2 (CaO)	0 and 10	1700	[6.26]
ThO_2 (Y_2O_3)	0 and 20	1700	[6.27]
ZrO_2 (CaO)	14 and 20	1500	[6.28]
ZrO_2 (Y_2O_3)	7 and 50	1500	[6.29]
HfO_2 (CaO)	14 and 20	1800	[6.30]
HfO_2 (Y_2O_3)	11 and 57	1800	[6.31]

[a]   Composition in mol-% of either CaO or Y_2O_3.
[b]   The difference in the phase boundaries reported by various investigatlrs are usually between 1 and 3 mol-%. [6.5].

Phase boundaries for the fluorite single phase in HfO_2(CaO) and HfO_2(Y_2O_3) solid solutions at 1800 °C are also tabulated in Table 6.1. Although one phase diagram for the HfO_2(CaO) system shows the fluorite phase decomposing below 1450 °C [6.30], conductivity measurements with HfO_2(CaO) solid solutions indicate that the fluorite phase is stable down to at least 800 °C [6.36]. The fluorite phase in the HfO_2(CaO) system is reported to be stable at temperatures lower than 1000 °C [6.31].

Comparison of the fluorite phase boundaries for the oxide solid solutions tabulated in Table 6.1 indicates that the Y_2O_3 stabilized solid solutions have wider compositional ranges than those of the CaO stabilized ones. The similarity between the fluorite structure and the cubic, C-type rare-earth oxide structure of Y_2O_3 may account for the wider compositional ranges. Calcium oxide has the halite (NaCl) crystal structure.

6.2 Defects and Conduction Mechanisms in Fluorite-Type Oxides

6.2.1 Defects in Predominately Ionic Solids

It is helpful to review briefly the type of defects in ionic solids before describing conduction mechanisms in fluorite-type oxides. The entropy contribution to the Gibbs free energy requires the presence of a finite concentration of point defects in solids at any temperature above 0 K [6.37, 38, p. 263]. Even an exactly stoichiometric MO oxide contains point defects. However, to preserve the lattice-site ratio and electrical neutrality, a positively charged ionic defect will be compensated by the presence of a negatively charged ionic defect. The most common defect structures in ionic solids are vacancy pairs (Schottky defects) and vacancy-interstitial pairs (Frenkel defects) [6.22, 39].

Non-stoichiometry occurs when there is an excess concentration of one type of ionic defect over that occurring at the stoichiometric composition. However, the stoichiometric ratio of cationic to anionic lattice *sites* remains constant even when the composition of the compound changes. Thus any change in concentration of an anionic defect must be accompanied by a corresponding change in the number of electronic defects to preserve electroneutrality.

Electronic defects can be formed by thermal excitation of electrons from the valence band to the conduction band. Equilibrium between free electrons (e') in the conduction band and electron holes $(h\dot{})$ in the valence band can be expressed as [6.40, p. 501]

$$0 \rightleftarrows e' + h\dot{} \tag{6.1}$$

$$K_1 = np \propto \exp(-E_g/kT) \tag{6.2}$$

where n and p denote the concentrations of electrons and electron holes, respectively, K_1 is the equilibrium constant and E_g is the energy difference between the valence and the conduction bands (bandgap energy). In this chapter the Kröger-Vink defect notation [6.39, 41] is used in expressing defect equilibria. In this notation the symbol indicates the defect, the subscript the location of the defect and the superscripts $'$ and $\dot{}$ denote negative and positive effective charges, respectively. The effective charge of a defect is the charge on the defect relative to the perfect crystal. For example, removing an oxygen ion, $O^=$, from an oxide lattice creates an oxygen vacancy, which is indicated as $V_O^{\cdot\cdot}$ because oxygen and two electrons are removed from the crystal.

Concentrations of ionic and electronic defects are important because electrical charge is transported by movement of these defects. The electrical conductivity (σ) of a solid is related to its defect concentration by

$$\sigma = \sum_i c_i q_i \mu_i + nq_e \mu_e + pq_h \mu_h \tag{6.3}$$

where c is the ionic defect concentration, q is the charge, μ is the mobility (i.e., the mean particle velocity per unit potential gradient), and the subscripts i, e, and h denote ions, electrons and electron holes, respectively.

Experimental values for the mobilities of electronic defects in ionic compounds are usually 100 to 1000 times greater than those of the ionic defects [6.7, p. 131, 42]. Thus ionic conduction should predominate only in those solids in which the concentration of electronic defects is less than the ionic defect concentration by a factor of 1000 or more. The larger the bandgap energy, the smaller is the value of K_1 in (6.2) and the lower is the concentration of electronic defects. Compounds which exhibit predominately ionic conductivity usually have a bandgap energy greater than 3 electron volts (eV).

Intrinsic defects are fixed by thermodynamic equilibria in the pure compound, while extrinsic defects are established by the presence of aliovalent impurities. When the concentrations of intrinsic defects are small, the conductivities of ionic compounds are very sensitive to impurity effects. Aliovalent impurities can significantly increase the ionic conductivity, particularly at low temperatures [6.42]. To maintain electroneutrality, a soluble aliovalent ion in an ionic compound will be compensated by an increase in the concentration of an ionic defect.

The effect of impurities on the concentrations of electronic defects is not well understood. One qualitative guide is that variable valence impurity ions like iron or manganese usually increase the electronic defect concentration. Such variable-valence ions can either donate electrons to the conduction band or accept electrons from the valence band.

6.2.2 Defect Equilibria and Conduction Mechanisms in ThO$_2$

The relationship between defect equilibria and electrical conductivity in fluorite-type oxides can be described using ThO$_2$ as an example. Many studies of oxides, fluorides and oxyfluorides having the fluorite lattice have established that the ionic defects occurring in this structure are interstitial anions and anionic vacancies [6.43]. Thus the predominant intrinsic ionic defects in thoria are believed to be oxygen vacancies, $V_O^{\cdot\cdot}$, and oxygen interstitials, O_i'' [6.39, 41, 43–46].

With these ionic defects, electroneutrality in pure ThO$_2$ requires that

$$p+2[V_O^{\cdot\cdot}]=n+2[O_i''] \tag{6.4}$$

where the brackets in (6.4) represent concentrations of the ionic defects. Equilibrium between electrons and electron holes is represented by (6.1) and (6.2). Considering only fully ionized defects, the formation of an anionic Frenkel defect in ThO$_2$ can be written

$$O_O \rightleftarrows V_O^{\cdot\cdot}+O_i'' \tag{6.5}$$

$$K_2=[V_O^{\cdot\cdot}][O_i''] \tag{6.6}$$

where O_O represents oxygen in an anionic lattice site. When the concentration of oxygen anions is large compared to the number of defects, the simple mass action law represented by (6.6) can be used.

Equations (6.5) and (6.6) are valid at a constant temperature and composition. According to the phase rule, any variation in oxygen pressure in equilibrium with ThO$_2$ changes the concentrations of ionic and electronic defects. At high oxygen pressures the predominant defects are interstitial anions and electron holes; thus the defect equilibria can be expressed as

$$\tfrac{1}{2}O_2(g)\rightleftarrows O_i''+2h^{\cdot} \tag{6.7}$$

$$K_3=[O_i'']p^2/P_{O_2}^{1/2} . \tag{6.8}$$

As the equilibrium oxygen pressure is increased, the electron hole concentration increases to preserve electroneutrality as required by (6.4). If $[O_i'']$ is large and essentially constant, (6.8) reduces to

$$p\propto P_{O_2}^{1/4} . \tag{6.9}$$

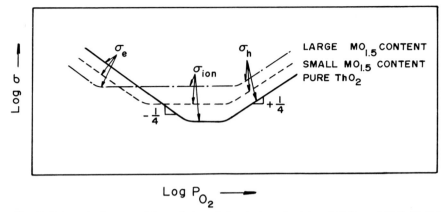

Log σ

Log P_{O_2}

Fig. 6.2. Schematic diagram of the variation of electrical conductivity of ThO_2 and $ThO_2(Y_2O_3)$ solid solutions with oxygen pressure at constant temperature

Assuming that the electron-hole mobility does not vary with its concentration, electron-hole or p-type conductivity should be directly proportional to $P_{O_2}^{1/4}$ in the high oxygen pressure range where (6.9) is valid.

At very low oxygen pressures, the predominant defects in equilibrated ThO_2 will be oxygen vacancies and electrons; thus the defect equilibrium is

$$O_O \rightleftarrows 1/2 O_2(g) + V_O^{\cdot\cdot} + 2e' \tag{6.10}$$

$$K_3 = [V_O^{\cdot\cdot}]n^2 P_{O_2}^{1/2}. \tag{6.11}$$

If $[V_O^{\cdot\cdot}]$ is large and essentially constant, (6.11) reduces to

$$n \propto P_{O_2}^{-1/4}. \tag{6.12}$$

Assuming that the electron mobilities are constant, n-type electronic conductivity should be directly proportional to $P_{O_2}^{-1/4}$ in this low oxygen pressure region.

In an "intermediate" oxygen pressure region, there may be a negligible effect on the concentration of electronic and ionic defects from those established by (6.2) and (6.6). If the defect mobilities are constant, the electrical conductivity is independent of oxygen pressure in this intermediate region. The variation of the electrical conductivity of ThO_2 with oxygen pressure at constant temperature is schematically represented by the solid line in Fig. 6.2. Such a diagram should have three regions: a low P_{O_2} region where $\sigma \approx \sigma_e \propto P_{O_2}^{-1/4}$, an intermediate P_{O_2} region where σ is independent of P_{O_2}, and a high P_{O_2} region where $\sigma \approx \sigma_h \propto P_{O_2}^{-1/4}$.

The high temperature electrical conductivity of ThO_2 has been measured by four investigators [6.47–50]. Results from three studies at 1000 °C are shown in Fig. 6.3. Although only total conductance data are given in the fourth study [6.50], excellent agreement is reported with the conductivity results of *Lasker*

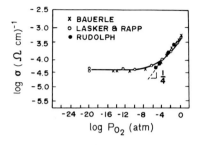

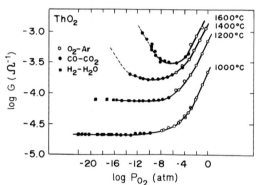

Fig. 6.3. Variation of the electrical conductivity of ThO_2 with oxygen pressure at 1000 °C. Results are from *Rudolph* [6.47], *Bauerle* [6.48] and *Lasker* and *Rapp* [6.49]

Fig. 6.4. Total conductance of ThO_2 as a function of oxygen pressure [6.50]

and *Rapp* [6.49] at 1000 °C. As shown in Fig. 6.3, *p*-type electronic conductivity becomes significant in ThO_2 when the oxygen pressure is greater than 10^{-6} atm at 1000 °C.

At oxygen pressures below 10^{-8} atm, the conductivity of ThO_2 is independent of P_{O_2}. The independence of oxygen pressure is a necessary, but not sufficient, condition for ionic conduction. An electronic conductor in which the concentrations of electrons and holes are established by (6.1) would also exhibit conductivity independent of oxygen-pressure. However, *Lasker* and *Rapp* [6.49] have demonstrated that the conductivity of ThO_2 below 10^{-8} atm is ionic by ancillary galvanic-cell experiments. The results shown in Fig. 6.3 indicate that no *n*-type electronic conductivity is observable in ThO_2 even at oxygen pressures as low as 10^{-20} atm at 1000 °C. Electrical conductivity data for ThO_2 at temperatures between 800 and 1100 °C also show behavior similar to that illustrated in Fig. 6.3 [6.49]. The total conductance data shown in Fig. 6.4 [6.50] indicate that *n*-type conductivity becomes significant in ThO_2 at temperatures above 1400 °C.

The 1/4 power dependencies in (6.9) and (6.12) are obtained by assuming that the concentrations of ionic defects in ThO_2 is much greater than that of the electronic defects. In high purity ThO_2, this is true if K_2 of (6.6) is much greater than K_1 of (6.2). However, the ThO_2 used by *Bauerle* [6.48] and by *Lasker* and *Rapp* [6.49] contained 200–400 ppm of trivalent impurities. For every two trivalent ions added to ThO_2, one oxygen vacancy is formed to preserve electroneutrality, and the number of oxygen vacancies is increased. *Bauerle*

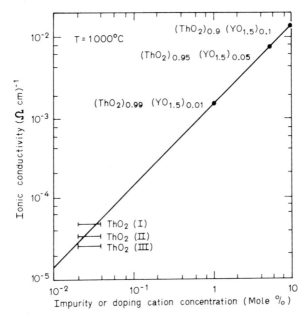

Fig. 6.5. Variation of the ionic conductivity of ThO₂ with the concentration of Y₂O₃ and of trivalent impurities (0.02–0.04%) at 1000 °C [6.48]

[6.48] has used Fig. 6.5 to demonstrate that the ionic conductivity of ThO_2 is due to such impurity effects. The ionic conductivities of three different $ThO_2(Y_2O_3)$ solid solutions are plotted as a function of cation concentration in the upper right-hand corner of Fig. 6.5. The trivalent cation impurity content of the "pure" ThO_2 is between 0.02 and 0.04 mol-%. As seen in Fig. 6.5, the ionic conductivities of three "pure" ThO_2 samples fall on the line extrapolated from the solid-solution results. *Rapp* [6.51] also concludes that his earlier ionic conductivity results for ThO_2 [6.49] were influenced by cationic impurities. There is another indication that observed ionic conductivities at oxygen pressures below 10^{-6} atm in ThO_2 are extrinsic. The activation energy of 0.9 eV estimated for ionic conductivity in ThO_2 [6.50] is close to the value reported for the migration enthalpy for oxygen in thoria-yttria solid solutions [6.49, 52]. If ionic conductivity in ThO_2 were due to the migration of intrinsic defects, the activation energy would include the formation enthalpy of the oxygen vacancies and would be much larger than the oxygen migration enthalpy.

6.3 Ionic Conduction in Oxide Solid Solutions

6.3.1 ThO₂(Y₂O₃) Solid Solutions

Pycnometrically measured densities of $ThO_2(Y_2O_3)$ solid solutions agree with those calculated from an oxygen-vacancy defect model [6.27, 53–55]. The dissolution of yttria into the fluorite phase of ThO_2 can be written:

$$Y_2O_3 \rightarrow 2Y'_{Th} + 3O_O + V_O^{\cdot\cdot} \tag{6.13}$$

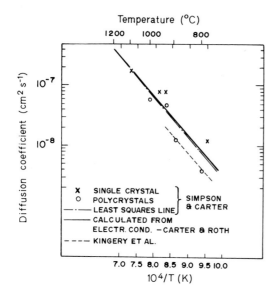

Fig. 6.6. Self-diffusion coefficient of oxygen in $ZrO_2(CaO)$ solid solutions as determined by isotope studies and by electrical conductivity. The diffusion data are those of *Kingery* et al. [6.56] for $Zr_{0.85}Ca_{0.15}O_{1.58}$ and those of *Simpson* and *Carter* [6.57] for $Zr_{0.858}Ca_{0.142}O_{1.858}$. The electrical conductivity data are those of *Carter* and *Roth* [6.60]

where (6.13) preserves the lattice-site ratio in ThO_2. Each addition of two yttrium ions creates one oxygen vacancy. At a fixed oxygen pressure, any increase in the oxygen-vacancy concentration must decrease the concentration of oxygen interstitials (6.6), increase the electron-hole concentration (6.8) and decrease the electron concentration (6.11).

Assuming that the conductivity attributable to a particular defect is directly proportional to its concentration, the effects of Y_2O_3 additions on the electrical conductivity of ThO_2 are schematically illustrated in Fig. 6.2. The ionic conductivity (σ_{ion}) and the electron-hole conductivity (σ_h) increase, while the electron conductivity (σ_e) decreases with increasing Y_2O_3 content. Experimental data for the electrical conductivity of $ThO_2(Y_2O_3)$ solid solutions at temperatures between 800 and 1100 °C and at oxygen pressures between 10^{-27} and 1 atm [6.49] follow the behavior schematically illustrated in Fig. 6.2.

6.3.2 Oxygen Diffusion and Ionic Conduction

As indicated in Fig. 6.5, the ionic conductivity in $ThO_2(Y_2O_3)$ solid solutions increases as the oxygen-vacancy concentration increases. However, this is rather indirect evidence that ionic conduction is due to oxygen-ion migration. More direct evidence is available for $ZrO_2(CaO)$ electrolytes in which the self-diffusion coefficients of oxygen, zirconium and calcium have been measured [6.56–59]. Results of these radioactive-tracer measurements on polycrystalline samples are tabulated in Table 6.2. Oxygen self-diffusion is at least 10^5 to 10^6 times faster than that of the cations.

The self-diffusion coefficient of oxygen in $ZrO_2(CaO)$ electrolytes has been calculated from ionic conductivity data [6.60] using the Nernst-Einstein

Table 6.2. Diffusion coefficients (D) in ZrO_2 (CaO) electrolytes at 1000 °C

Mol-% CaO	Ion	D [cm^2/s]	AE^a [Kcal/mole]	Reference
15	O^{-2}	$4.2 \cdot 10^{-8}$	30.4	[6.56]
14.2	O^{-2}	$7.9 \cdot 10^{-8}$	31.2	[6.57]
12, 16	Zr^{+4}	$4.6 \cdot 10^{-18b}$	92.5	[6.58]
16	Ca^{+2}	$2.8 \cdot 10^{-18b}$	100.2	[6.58]
15	Zr^{+4}	$1.2 \cdot 10^{-13c}$	61.8	[6.59]
15	Ca^{+2}	$4.4 \cdot 10^{-14c}$	98.8	[6.59]

^{a} Activation energy.
^{b} Extrapolated from 1700 °C.
^{c} Extrapolated from 1550 °C.

equation [6.22, p. 161]. As shown in Fig. 6.6, the agreement between the calculated and measured values for the oxygen self-diffusion coefficients is excellent. The results shown in Table 6.2 and Fig. 6.6 confirm that ZrO_2(CaO) solid solutions are ionic conductors due to oxygen-ion migration.

6.3.3 Effects of Composition

As shown in the upper right-hand corner of Fig. 6.5 the ionic conductivity in dilute $ThO_2(Y_2O_3)$ solid solutions is directly proportional to the yttria concentration in the range of 0.01 to 0.1 mol-% $YO_{1.5}$ [6.48]. This ideal behavior occurs in dilute solutions when interactions between defects are minimal. Under these conditions, the simple mass-action-law approach represented by (6.2, 6, 8, 11) is valid. However, in more concentrated solid solutions, such an approach is no longer realistic because defect interactions can be expected.

The variation of ionic conductivity with oxygen vacancies in concentrated $ThO_2(Y_2O_3)$ solid solutions at 1000 °C is shown in Fig. 6.7. The results have been obtained using polycrystalline electrolytes prepared by sintering fine powders. Small density differences in the sintered samples could account for the differences among the three investigations [6.48, 49, 52], which are shown in Fig. 6.7. Errors introduced by the presence of grain boundaries are believed to be insignificant compared to other experimental uncertainties [6.5]. Essentially identical conductivity data for single-crystal and polycrystalline ZrO_2(CaO) electrolytes have been obtained after correcting for 3–6% porosity in the latter [6.60]. The purity level of polycrystalline samples can be extremely important, because impurities usually segregate to the grain boundaries. Thus grain boundary effects have been observed in $ZrO_2(Y_2O_3)$ samples having higher impurity levels [6.61, 62].

In Fig. 6.7 the ionic conductivity in $ThO_2(Y_2O_3)$ solid solutions exhibits a maximum value around 3 to 4% oxygen vacancies and decreases at higher vacancy concentrations. Similar maxima are observed in ZrO_2(CaO) and $ZrO_2(Y_2O_3)$ solid solutions [6.5, 42]. The variation of the ionic conductivity of ZrO_2(CaO) solid solutions with composition and temperature is shown in Fig. 6.8 [6.60]. At 1000 °C, the maximum value ($\sim 5.5 \cdot 10^{-2}$ (Ω cm)$^{-1}$) occurs

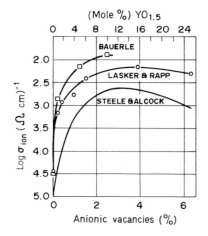

Fig. 6.7. Ionic conductivity of $ThO_2(Y_2O_3)$ solid solutions as a function of oxygen-vacancy concentration (mol-% $YO_{1.5}$) at 1000 °C. Results are from *Bauerle* [6.48], *Lasker* and *Rapp* [6.49], and *Steele* and *Alcock* [6.52]

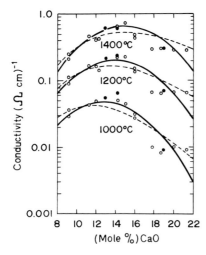

Fig. 6.8. Ionic conductivity of $ZrO_2(CaO)$ solid solutions as a function of composition and temperature [6.60]. Data for single-crystal (solid circles) and polycrystalline samples (open circles) are shown. The solid curves are calculated for the "zone" model; the dashed curves are from the "cluster" model

between 13 and 14 mol-% CaO. These results are in good agreement with earlier measurements summarized by *Etsell* and *Flengas* [6.5].

The interpretation of the conductivity data shown in Fig. 6.8 is complicated by the observation of an ordering process in $ZrO_2(CaO)$ solid solutions near 1000 °C, which can be reversed by heating for short times at 1400 °C [6.60]. While a decrease in the ionic conductivity is observed during long annealing (aging) experiments at 1000 °C (2200 h for polycrystalline specimens and 1080 h for single crystals), ordering does not appreciably change the activation energy for conduction [6.60, 63–65]. The ordering process has been investigated using dielectric [6.60] and mechanical [6.65] relaxation techniques.

Interactions between calcium cations, Ca''_{Zr}, and oxygen vacancies, $V^{..}_O$, are involved in this order-disorder process, and the internal friction in $ZrO_2(CaO)$ solid solutions has been explained in terms of the formation of neutral

$(Ca''_{Zr} - V_O^{\cdot\cdot})$ pairs [6.65]. Because this explanation could not account for their observed shift in the dielectric loss peak with aging, *Carter* and *Roth* [6.60] have introduced a "cluster" model and a "zone" model to interpret the variation of conductivity with composition.

In the "cluster" model, the defects are assumed to interact to form clusters. The pre-exponential term in the temperature-dependent equations for conductivity varies with composition according to x^n where x is the mole fraction and n is a constant related to the activation entropy of the cluster. When n is given the value of 7, good agreement with experimental conductivity data is obtained, as is shown by the dotted lines in Fig. 6.8. Classical ionic-conductivity models require a value of 1 or less for n, and the extraordinary strong concentration dependence implied by a value of 7 suggests that ionic transport involves a network or other multiplicative process.

As shown by the solid lines in Fig. 6.8 the conductivity data also fit a Poisson-type distribution relationship, which suggests an interpretation based on compositional fluctuations in the disordered phase. This "zone" model implies that an interlocking network of coherent phases of different conductivities exists in $ZrO_2(CaO)$ solid solutions. This picture is supported by x-ray, neutron diffraction, and electron microscope studies and by dielectric loss data [6.60].

The observed decreases in ionic conductivity with an essentially constant activation energy during aging experiments at $1000\,°C$ are also attributed to ordering of the "zones". *Carter* and *Roth* [6.60] suggest that only the pre-exponential term in the temperature-dependent equation for conductivity should be affected by changes in microstructure and that the activation energies for conduction in the ordered and disordered state should be the same.

The "cluster" and "zone" models provide plausible explanations for the compositional and temperature dependence of the conductivity in $ZrO_2(CaO)$ solid solutions. These models are helpful in interpreting experimental data until more quantitative treatments of oxygen-ion transport in concentrated oxide solid solutions are available.

6.3.4 Effects of Temperature and Oxygen Pressure

Most applications of solid electrolytes require that electronic conduction be negligible. For practical purposes, this requirement is usually fulfilled by any electrolyte in which the electronic conductivity is less than 1% of the total conductivity. The temperature and oxygen pressure region over which the electronic contribution is less than 1% is called the electrolytic domain. Electrolytic domains for the $ThO_2(Y_2O_3)$ and $ZrO_2(CaO)$ electrolytes are shown in Fig. 6.9 [6.66]. At all temperatures and oxygen pressures inside the boundary lines shown in Fig. 6.9, the ionic fraction of the total conductivity is greater than 0.99. For example, the electrolytic domain for $ZrO_2(CaO)$ is between $\sim 10^5$ and $\sim 10^{-18}$ atm P_{O_2} at $1000\,°C$. At oxygen pressures above $\sim 10^5$ atm, p-type electronic conductivity is greater than 1%, while n-type conductivity becomes greater than 1% at oxygen pressures below $\sim 10^{-18}$ atm.

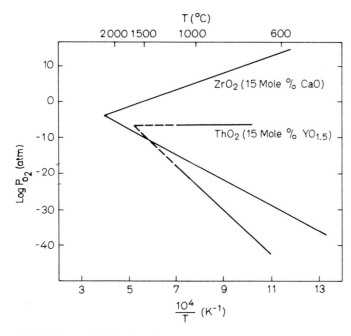

Fig. 6.9. Electrolytic domains for $ZrO_2(CaO)$ and $ThO_2(Y_2O_3)$ electrolytes [6.66]

The electrolytic domain for the $ThO_2(Y_2O_3)$ electrolyte is between $\sim 10^{-6}$ and $\sim 10^{-25}$ atm oxygen pressure at 1000 °C.

As illustrated in Fig. 6.9, the width of the electrolytic domains decreases with increasing temperature. This decrease is an indication that the activation energy for electronic conductivity is larger than that for ionic conductivity in oxide electrolytes. Because oxygen-vacancy concentrations are high, activation energies for ionic conduction in oxide solid solutions are equal to the oxygen-ion migration enthalpies, which are between 1.1 and 1.3 eV in $ThO_2(Y_2O_3)$ [6.49, 50] and in $ZrO_2(CaO)$ [6.56, 60]. The activation energy for electronic conduction is equal to the bandgap energy, which must be very high because the number of electronic defects in these oxide electrolytes is extremely small. For example, the bandgap energy for a $ZrO_2(CaO)$ solid solution is estimated to be ~ 5.6 eV [6.22, p. 164].

The exact location of the boundary lines shown in Fig. 6.9 is strongly influenced by the impurity content of the electrolyte. Variable-valence impurity ions like iron and manganese are particularly harmful, for they can donate or accept electrons by changing their valence state. Assuming that electronic mobilities are 1000 times higher than ionic mobilities, only 60 ppm (mole) of an impurity (which is compensated for by excess electrons) introduces 1 % electronic conductivity into an oxide electrolyte with 3 mol-% oxygen vacancies [6.67]. Thus it is not surprising that various investigators differ on the exact location of the domain boundaries [6.66]. For example, there is ample evidence that the lower boundary line for $ThO_2(Y_2O_3)$ should be at lower oxygen pressures than that shown in Fig. 6.9 [6.12, 49, 68].

6.4 Applications of Oxide Solid Electrolytes

Most applications of oxide electrolytes involve the use of zirconia- and thoria-based solid solutions. These two electrolyte systems differ in their ionic conductivities and applications. Comparison of Figs. 6.7 and 6.8 indicates that zirconia-based electrolytes have higher ionic conductivities. Thus zirconia-based electrolytes are used in high temperature fuel cells [6.20]. From the electrolytic domains illustrated in Fig. 6.9, it is apparent that thoria-yttria electrolytes exhibit significant p-type electronic conductivity at oxygen pressures above 10^{-6} atm and that zirconia-calcia electrolytes exhibit n-type electronic conductivity at oxygen pressure below 10^{-22} atm at 1000 °C. Thus the $ThO_2(Y_2O_3)$ electrolytes cannot be used in high P_{O_2} environments such as air, and zirconia electrolytes exhibit electronic conductivity in highly reducing atmospheres.

6.4.1 High Temperature Research

Since 1957 zirconia-based and thoria-based electrolytes have been used widely in equilibria investigations and kinetic studies at elevated temperatures. In this section, the electrochemical principles and technique necessary for the successful use of oxide electrolytes in high-temperature research are summarized.

Equilibrium or Thermodynamic Investigations

Wagner [6.69] has derived the relation between the electromotive force (E) of an electrochemical cell and the chemical potential (μ_x) at each electrode:

$$E = \frac{-1}{|Z_x|F} \int_{\mu_x''}^{\mu_x'} t_{ion} d\mu_x \qquad (6.14)$$

where F is Faraday's constant, $|Z_x|$ is the absolute value of the valence of the X ion in the electrolyte, μ_x' is $<\mu_x''$, and t_{ion} is the ionic transference number, which is the fraction of current carried by ionic carriers. For an oxide electrolyte, $t_O =$ is 1, and (6.14) reduces to

$$\Delta G = \mu_0' - \mu_0'' = -2FE. \qquad (6.15)$$

Whether the cation, the anion or both are mobile in the oxide electrolyte, (6.15) is valid when t_{ion} is >0.99. Because a solid exhibits predominantly ionic conduction only over a finite partial pressure range of a component, it is necessary to know the electrolytic domain boundaries as shown in Fig. 6.9.

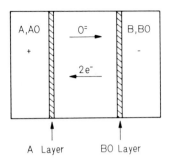

A Layer   BO Layer

Fig. 6.10. Schematic diagram of an electrochemical cell in which there is a small amount of electronic conductivity in the oxide electrolyte

If t_{ion} is less than unity, (6.14) can be integrated when the variation of t_{ion} with μ_x is known [6.70]. However, there can be experimental difficulties when $t_{ion} < 0.99$, particularly with solid electrodes. With any electronic current, electrons will migrate from the negative to the positive electrode, and oxygen ions will move in the opposite direction to maintain local electroneutrality throughout the oxide electrolyte. If diffusional processes cannot supply or remove oxygen rapidly enough at the electrode-electrolyte interfaces, there can be oxide formation at the negative electrode and metal formation at the positive electrode. This is illustrated schematically in Fig. 6.10 for the two-phase metal-metal oxide electrodes. Such layer formations will cause unstable emfs because the oxygen potential is then not fixed at the electrode-electrolyte interfaces.

Even when layer formation does not occur, the chemical potential of oxygen can differ from that established by the metal-metal oxide equilibrium. For example, recent measurements with various metal-metal oxide electrodes [6.71–73] have demonstrated that a steady-state overvoltage is obtained when extremely small currents (1–$100\,\mu A$) are passed through symmetrical electrochemical cells of the type

$$M, MO/ZrO_2(CaO)/M, MO. \tag{6.16}$$

The overvoltage arises from steady-state oxygen concentration gradients in the metal electrodes. Results indicate that the major oxygen transfer at the electrode-electrolyte interface occurs between the electrolyte and the metal particles in the electrode [6.71–73]. Results shown in Fig. 6.11 for Ni–NiO, Fe–FeO, and Cu–Cu$_2$O electrodes illustrate the variation of the steady-state overvoltage (η) with the imposed current (I) at $900\,°C$. One important practical conclusion from Fig. 6.11 is that Cu–Cu$_2$O electrodes can adjust to small cell current much more easily than either Fe–FeO or Ni–NiO electrodes.

Electrochemical cells with oxide solid electrolytes have been used to measure the chemical potentials of oxygen in oxides and in metals, the thermodynamic stabilities of oxides and intermetallic compounds, and the thermodynamic properties of alloys. Such equilibrium measurements have been extensively reviewed [6.5, 7, 74–76]. *Electrochemical cell measurements are deceptively simple.* The emf is measured by a null method in which no current is passed across the cell. However, there are several experimental problems which limit the

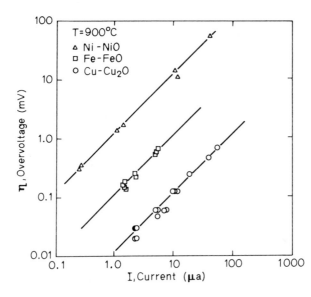

Fig. 6.11. Overvoltage versus current for cell (6.16) at 900 °C using Cu–Cu₂O, Fe–FeO and Ni–NiO electrodes [6.71, 72]

successful use of electrochemical cells in high temperature equilibrium measurements [6.67].

Experimental problems with the electrolyte include electronic conductivity, porosity and microcracks. As described previously, a small electronic current in the electrolyte can change the chemical potentials of oxygen at the two electrode-electrolyte interfaces. When this occurs, the cell emf decreases continuously. To avoid this problem, one must use high purity electrolytes over a temperature and oxygen pressure range within the electrolytic domain. It should be noted that $ThO_2(Y_2O_3)$ electrolytes have been used successfully with electrodes having equilibrium oxygen pressures as low as 10^{-25} atm at 1000 °C [6.77–80]. Porosity or microcracks in the electrolyte can provide paths for the migration of molecular oxygen from one electrode to the other, and thus change the oxygen chemical potentials at the two electrode-electrolyte interfaces. With proper preparation and sintering techniques, porosity in the electrolyte will not be a problem. If a cell is heated or cooled too rapidly, thermal stresses cause microcracks to form in the electrolyte. This can be a particularly severe problem when using electrolyte tubes.

Difficulties in maintaining a fixed chemical potential at the electrode-electrolyte interface can be caused by the influence of the gaseous phase, undesirable chemical reactions and deviations from equilibrium among electrode constituents [6.67]. Electrochemical cell measurements are usually made in a flowing inert gas (argon) atmosphere which is first passed through a purification furnace containing copper or titanium gettering chips. When measuring extremely low oxygen potentials, it is necessary to place an oxygen getter (tantalum or niobium foil) inside the reaction tube upstream from the cell, to remove any residual oxygen. By varying the flow rate of the inert gas, one can determine if the cell emf is independent of flow rate. If there is a flow rate effect, oxygen in the inert gas may be establishing a mixed potential at the electrode-electrolyte interface. To minimize this problem, one can improve the interfacial contact by using electrodes and electrolytes with prepolished smooth surfaces and by using springs or weights to hold the cell tightly together. In some electrochemical cells, the difference in the oxygen chemical potential between the two electrodes may be so large that the inert gas atmosphere will tend either to reduce one electrode or to oxidize the other. To avoid this, one should choose a reference electrode which establishes an oxygen pressure within a factor of 10^5 atm of the estimated P_{O_2} of the unknown electrode. A useful list of metal-metal oxide reference electrodes has been tabulated [6.7]. If a suitable reference electrode is unavailable, it is useful to separate the electrodes into two compartments, either by using an electrolyte tube or by sealing a refractory tube to an electrolyte pellet [6.75].

Any chemical reaction within the electrode or between the electrode and the electrolyte can lead to chemical potential variations in the electrode. When using binary alloy-oxide electrodes, one can prevent an alloy-oxide reaction by using only those alloys in which the oxide of one constituent metal is at least 20 kcal/mole more stable than that of the other [6.81, 82]. For this reason the Ta–V system cannot be investigated using an oxide electrolyte, although electrochemical cell measurements of the Ta–Mo and Ta–W systems have been successful [6.78, 79]. Reactions between the electrolyte and any electrode constituent also may occur. Because of the high thermodynamic stabilities of ZrO_2 and ThO_2, such reactions will occur only when very stable ternary oxide compounds, intermetallic phases or solid solutions are formed as reaction products. For example, a $YFeO_3$ phase resulting from a reaction between a $ThO_2(Y_2O_3)$ electrolyte and an Fe–FeO electrode has been detected by x-ray diffraction of the electrolyte surface [6.77].

With careful cell preparation and design, problems with the electrolyte, the gaseous atmosphere and undesirable chemical reactions can be eliminated. However, the equilibration of the electrode constituents is always of concern. Equilibration times are limited by diffusional processes and will vary with the particular electrode [6.67]. With two- and three-phase electrodes, pressed electrode pellets of small particles (-325 mesh) can decrease diffusion distances and promote interphase contact. With solid electrodes, it is usually necessary to equilibrate the electrode mixture at a temperature within the experimental range before assembling the electrochemical cell [6.83, 84].

Even the most carefully designed electrochemical cells are limited to a finite temperature range. With oxide-solid electrolytes, experiments have ranged from 500° [6.85, 86] to 1600 °C [6.12], but most are in the 700° to 1200 °C range. The low temperature limit is usually established by equilibration difficulties among the electrode constituents because of slow diffusional processes. However, at very low temperatures the conductivity of the electrolyte could be lower than that necessary to establish a stable emf across the cell. The high temperature limit is determined either by the appearance of electronic conductivity or by undesirable chemical reactions.

Kinetic Studies

Kinetic measurements using solid electrolytes are based on Faraday's law, which relates the rate, j_i (moles/s), at which an ionic species of valence z_i moves through the electrolyte to the current (I) passed through the cell:

$$j_i = It_i/z_iF \tag{6.17}$$

where t_i is the transference number for ionic species i. It should be emphasized that there is only one mobile ion in most solid electrolytes. This is in contrast to liquid electrolytes, in which a number of ionic species are involved in Faradaic processes at the electrode-electrolyte interface. Thus kinetic studies using solid electrolytes usually are better defined and easier to interpret than those involving liquid electrolytes. Another advantage of using electrochemical cells in kinetic measurements is that the reaction rate and the chemical driving force can be measured simultaneously with high precision.

Comprehensive reviews of the use of solid electrolytes in kinetic measurements have been published [6.7, 76, 81, 87–89]. Kinetic studies using solid-oxide electrolytes include coulometric titrations of oxygen [6.85, 86, 90–98], measurements of the oxygen diffusivity in liquid and solid metals [6.73, 99–107], and determinations of phase boundary reaction rates [6.108–111]. Coulometric titration is not strictly a kinetic measurement but a technique for varying the composition of an electrode by passing a given amount of current through an electrochemical cell. However, the number of equivalents of charge passed through the cell is related by (6.17) to the number of moles of oxygen which enter or leave the electrode. After the titration, sufficient time must be allowed for the electrode to equilibrate. For a reasonable equilibration time, the electrode must have either a rapid diffusion coefficient or be sufficiently thin. These parameters can be estimated using $\tau \approx x^2/D$, where τ is the equilibration time, x the sample thickness and D the diffusivity. For a 1 mm thick electrode pellet and $D \approx 10^{-6}$ cm^2/s, the equilibration time is 10^4 s (≈ 3 h). Coulometric titrations of oxygen in liquid lead [6.85] and tin [6.86] have been carried out between 500° and 750 °C. Coulometric titration also has been used to vary the composition of the following oxide phases: NiO [6.90], FeO [6.91–94], UO_2 [6.95, 96], TiO_2 [6.97], and NbO_2 [6.97].

The coulometric titration technique using solid electrolytes has been reviewed previously [6.7, 87, 98], so only a few major experimental considerations will be discussed in this section. For valid experiments with oxide electrolytes, all oxygen from the electrolyte must enter the electrode, and there can be no loss to either the gas phase or the lead material. To prevent harmful interactions with the gaseous atmosphere, successful experimental designs [6.90, 92, 97] incorporate the electrode into a separate compartment which is sealed from the cell atmosphere by viscous glass. *Rapp* and *Shores* [6.7] state that the glass seals are effective as long as the oxygen pressure in the electrode compartment is within a factor of 10^4 or 10^5 times that of the cell atmosphere. It is also essential to avoid polarization of the reference electrodes. Metal-metal oxide mixtures are the most commonly used reference electrodes, and polarization occurs because of the inability of the two-phase mixture to maintain a fixed oxygen chemical potential under high titration currents. The maximum tolerable current is different for each electrode. Results [6.71, 72], shown in Fig. 6.11, indicate that $Cu–Cu_2O$ is a better reference electrode than Fe–FeO, which is in turn better than Ni–NiO. Small amounts of electronic conduction can cause problems in fixing the electrode composition. A ZrO_2(15 % CaO) pellet having a 1-cm^2 cross section and 1 mm thickness has a resistance of 2.5 Ω at 1000 °C. For a cell emf of 500 mV, it was calculated that an electronic contribution of 1 % would allow a current of 2 mA to flow across the cell [6.98]. As shown in Fig. 6.10, oxygen ions will move in a direction opposite to that of the electrons, and the oxygen content of the electrode can change significantly.

Several authors have summarized diffusivity measurements using solid electrolytes [6.7, 87, 88, 89]. Using electrochemical cell techniques, the diffusivity of oxygen has been measured in liquid silver [6.99, 101–103, 111], copper [6.101, 103, 104], lead [6.105] and tin [6.107] and in solid silver [6.100, 102, 107], copper [6.73, 106] and nickel [6.107]. Potentiometric, potentiostatic and galvanostatic techniques were used in these measurements.

Potentiometric measurements of the diffusivity of oxygen in liquid silver [6.99, 111] and lead [6.105] have been made. The technique can be described by considering the schematic cell

$$A, AO/ZrO_2(CaO)/B(O), O_{2(g)} \tag{6.18}$$

where B(O) represents oxygen dissolved in metal B. If metal B is equilibrated with a particular oxygen pressure, a stable emf is measured for cell (6.18). Increasing the oxygen pressure above metal B will cause oxygen to diffuse into B. The gradual increase in the oxygen chemical potential at the electrode-electrolyte interface is followed by measuring the open circuit emf as a function of time. The oxygen diffusivity in metal B can be calculated from the time dependent emf if the diffusion geometry is known.

In a potentiostatic experiment, a constant emf is imposed across the cell, and the variation of current with time is measured. For example, the following cell was used to measure the oxygen diffusivity in solid silver [6.100].

$$Fe, FeO/ZrO_2(Y_2O_3)/Ag(O). \tag{6.19}$$

An emf is applied to cell (6.19) which establishes a negligible concentration of oxygen in silver at the $Ag–ZrO_2$ interface, and oxygen moves from the silver electrode through the electrolyte to the Fe, FeO electrode. From the measured cell current, the diffusivity of oxygen in the silver electrode can be calculated. In another nonsteady-state potentiostatic experiment [6.106] the oxygen diffusivity in solid copper was measured using the "double" cell:

$$FeO, Fe_3O_4/ZrO_2(CaO)/Cu(O)/ZrO_2(CaO)/FeO, Fe_3O_4 . \qquad (6.20)$$

An emf is imposed across the cell, and the copper electrode is equilibrated with a well-defined oxygen activity which is somewhat higher than that at the FeO–Fe_3O_4 electrodes. To start the kinetic experiment, the applied emf is increased quickly to a new constant value, and a new oxygen activity is established at both copper-electrolyte interfaces. Oxygen migration from the two interfaces into the interior of the copper is monitored by a continuous measurement of the cell current. From the solution of the appropriate diffusion equation and the time-dependence of the current, values for the solubility and diffusivity of oxygen in solid copper are obtained between 800° and 1000 °C.

In the galvanostatic technique, a constant current is imposed across the cell, and the variation of the emf with time is measured. A steady-state galvanostatic experiment has been used to measure the oxygen solubility-diffusivity product in solid copper, using the following electrochemical cell [6.73].

$$Cu, Cu_2O/Cu(O)/ZrO_2(CaO)/Cu(O)/Cu, Cu_2O \qquad (6.21)$$

where Cu(O) indicates a copper-foil electrode. After saturating the copper-foil electrodes with oxygen, a small constant current (1–25 μA) is passed across the cell. After an initial rise, the cell voltage reaches a steady-state value within 5 to 25 min depending on the temperature and the thickness of the copper-foil electrodes. The solubility-diffusivity product for oxygen in copper is calculated by combining the cell overvoltages with Fick's first law. Within experimental uncertainties, results from cells (6.20) and (6.21) are in good agreement.

To obtain reliable diffusion data using solid electrolytes, it is essential that intimate contact be achieved across the entire electrode-electrolyte interface. If interfacial contact is poor, molecular oxygen from the cell atmosphere can penetrate the metal-electrolyte interface. This is usually not a problem when the electrode is a liquid metal. With solid metal electrodes, the cells should be heated to a temperature just below the melting point of the metal and then annealed under pressure to obtain highly coherent electrode-electrolyte interfaces. The concentration polarization of reference electrodes described previously for coulometric titration cells can also be a problem in oxygen diffusion studies if the cell current is too high.

The use of solid electrolytes to investigate phase boundary reactions has been discussed by a number of reviewers [6.7, 70, 81, 87, 88]. Kinetic studies using gas electrodes and zirconia-based electrolytes [6.66, 109–110] have been fairly numerous because of their relevance to high temperature fuel cells. One example of a phase boundary study with an oxide electrode is the measurement of the oxidation rate of iron in CO_2/CO mixtures using a $ZrO_2(CaO)$ electrolyte [6.110]. A thin iron foil is sintered under pressure on the electrolyte. After equilibrating the iron foil with a particular CO_2/CO ratio, the CO_2/CO ratio is increased suddenly, and the cell emf is measured as a function of time. If the foil is thin enough, oxygen diffusion in the foil will be rapid, and the observed time dependence of the emf is a measure of the phase boundary reaction rate. Experimental difficulties include the formation of a nonporous coherent metal layer on the electrolyte and obtaining a thin enough layer so that solid-state diffusion effects are negligible.

6.4.2 Oxygen Probes or Sensors

Oxide solid electrolytes are used in many high temperature technical applications. They have been used to measure the oxygen partial pressures in gas streams [6.112–116], to remove oxygen impurities from gases [6.117] and to deoxidize induction-stirred copper melts [6.118]. One particularly active application of oxide solid electrolytes is their use as oxygen probes or sensors [6.17].

An oxygen probe is an electrochemical cell which is used to determine oxygen concentrations in gas mixtures or liquid metals. It consists of an oxide electrolyte, a reference electrode (a gas mixture or two-phase solid mixture having a known chemical potential of oxygen) and an unknown electrode. As indicated in (6.15), the chemical potential of the unknown electrode can be determined directly from the measured electromotive force of the cell.

Oxygen probes have been used to measure and monitor the oxygen content in liquid sodium [6.119–121], liquid copper [6.15, 16], and liquid steel [6.9–14, 122–124]. To illustrate the problems involved in the development of such probes, consider the three cells shown in Fig. 6.12. These three cells have been proposed as oxygen probes for use in liquid steel at 1500 to 1600 °C. These devices are used in a manner similar to a disposable immersion thermocouple. The probes are connected to a permanent receptacle and inserted directly into liquid steel. An electromotive force reading is usually obtained within 30 s, and the probes fail within 1 to 3 min after immersion.

Cell I in Fig. 6.12 uses a $ZrO_2(CaO)$ electrolyte tube with air or CO/CO_2 gas mixtures as the reference electrode [6.14, 122, 123]. Because of the poor thermal shock resistance of $ZrO_2(CaO)$ tubes, Cell I is not used commercially as an oxygen probe in liquid steel. *Fitterer* [6.11] has developed the design shown as Cell II, Fig. 6.12, in which a $ZrO_2(CaO)$ disc is sealed to a silica tube. Cell II can be used at liquid-steel temperatures because of the superior thermal shock resistances of silica tubes. One major difficulty with gas reference electrodes is the possibility of oxygen leakage through pores or microcracks in the electrolyte. Small pores and microcracks have been observed to form when impervious discs of $ZrO_2(CaO)$ were heated above 1400 °C [6.125]. Oxygen leakage through such macroscopic defects increases the oxygen concentration at the melt-electrolyte interface, which results in an erroneous cell emf.

Cell III [6.10–12, 124] in Fig. 6.12 uses a $Cr–Cr_2O_3$ reference electrode to minimize the oxygen-leakage difficulties of Cell II. The $Cr–Cr_2O_3$ reference

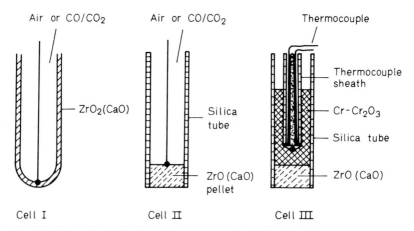

Fig. 6.12. Electrochemical cells used as oxygen probes in liquid steel

electrode establishes an oxygen chemical potential which is near that established at the melt-electrolyte interface. Thus there is less tendency for oxygen migration either through macroscopic defects (pores and microcracks) in the electrolyte or through imperfect seals around the electrolyte. Other metal-metal oxide mixtures such as $Mo–MoO_2$ can also be useful reference electrodes. Using Cell III with a zirconia-based electrolyte, oxygen contents as low as 10 ppm in liquid steel at 1600 °C have been measured [6.12]. When a thoria-based electrolyte is used, measurements of oxygen contents as low as 1 ppm in liquid steel are posssible [6.12].

6.4.3 High Temperature Fuel Cells

High temperature fuel cells are cells which operate above 300 °C. The electrolytes in these cells are usually molten salts or oxide solid solutions. Fuel cells using oxide electrolytes have many advantages such as simple construction, chemical inertness, and thermal and mechanical stability. However, oxide solid solutions have low ionic conductivities compared to molten salts. A $Na_2CO_3–K_2CO_3–Li_2CO_3$ molten salt has an ionic conductivity around 0.1 $(\Omega \, cm)^{-1}$ at 600 °C, while the ionic conductivities of $(ZrO_2)_{0.85}(CaO)_{0.15}$ and $(ZrO_2)_{0.91}(Y_2O_3)_{0.09}$ are ~ 0.06 and 0.1 $(\Omega \, cm)^{-1}$, respectively, at 1000 °C [6.5]. To minimize ohmic losses across oxide electrolytes, thin-film electrolytes must be fabricated, and fuel-cell temperatures must be above 900 °C. Fuel cells using $ZrO_2(CaO)$ or $ZrO_2(Y_2O_3)$ electrolytes are preferable to those using thoria-based electrolytes because zirconia-based solid solutions have higher ionic conductivities.

The use of zirconia-based electrolytes in high-temperature fuel cells has been extensively reviewed [6.5, 19, 20, 126]. The most commonly studied cell is:

$$Pt, H_2, H_2O/ZrO_2(CaO) \text{ or } ZrO_2(Y_2O_3)/O_2 \text{ or air, } Pt. \tag{6.22}$$

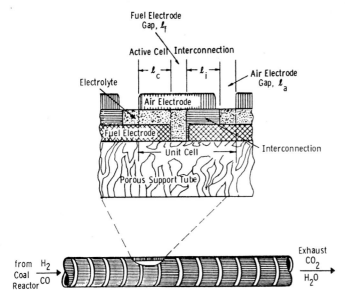

Fig. 6.13. Cross-sectional view of the Westinghouse thin-film fuel cell, which is constructed by the sequential deposition of 20–50 μm layers on a porous tube of stabilized zirconia [6.21]

Active research with oxide-electrolyte fuel cells occurred in the 1960's, but several practical problems have decreased the original enthusiasm. One severe problem has been the development of inexpensive materials to replace the platinum electrodes shown in cell (6.22).

A particularly novel development is the proposed use of oxide-electrolyte fuel cells in fluidized-bed coal reactors [6.21]. Heat released by the cells would be used in coal gasification, which would provide fuel gas for the fuel cell. The thermal efficiency of the coal-burning solid-electrolyte fuel-cell power system is estimated to be around 60%.

Figure 6.13 is an illustration of the cross sectioon of thin-film fuel cell designed for use in a coal reactor [6.21]. Cobalt-zirconia or nickel-zirconia cermets are used as fuel-electrode materials. Tin-doped indium oxide is the air electrode, and either calcia or yttria stabilized zirconia is the electrolyte. The interconnection material, which links an air electrode to the fuel electrode of an adjacent cell, must be a good electronic conductor and thermodynamically stable in both oxidizing (air electrode) and reducing (fuel electrode) environments. Manganese-doped cobalt chromite has been suggested as an interconnection material [6.127]. Active development of the fuel cell illustrated in Fig. 6.13 is currently underway at Westinghouse Research Laboratories.

Acknowledgement. The financial support of the U.S. National Science Foundation through Grant DMR 76-05779 and through the Laboratory for Research on the Structure of Matter at the University of Pennsylvania is gratefully acknowledged.

166   W. L. Worrell

References

6.1   W. Nernst: Z. Electrochem. **6**, 41 (1899)
6.2   E. Baur, H. Preis: Z. Electrochem. **43**, 727 (1937)
6.3   K. Kiukkola, C. Wagner: J. Electrochem. Soc. **104**, 308 (1957)
6.4   K. Kiukkola, C. Wagner: J. Electrochem. Soc. **104**, 379 (1957)
6.5   T. H. Etsell, S. N. Flengas: Chem. Rev. **70**, 339 (1970)
6.6   C. B. Alcock: *Electrochemical Force Measurements in High-Temperature Systems* (Institute of Mining and Metallurgy Pub., London 1968)
6.7   R. A. Rapp, D. A. Shores: *Physicochemical Measurements in Metals Research*, Part 2, ed. by R. A. Rapp (Wiley-Interscience, New York 1970) pp. 123–192
6.8   W. L. Worrell: Am. Ceram. Bull. **53**, 425 (1974)
6.9   R. Littlewood: Can. Met. Quart. **5**, 1 (1966)
6.10 E. T. Turkdogan, R. J. Fruehan: Yearbook Am. Iron Steel Inst. (1968), pp. 279–301
6.11 G. R. Fitterer: J. Metals **18**, 961 (1966); **19**, 92 (1967); **20**, 74 (1968)
6.12 R. J. Fruehan, L. J. Martonik, E. T. Turkdogan: Trans. Met. Soc. of AIME **245**, 1501 (1969)
6.13 E. Förster, H. Richter: Arch. Eisenhüttenw. **40**, 425 (1969)
6.14 C. Gatellier, K. Torssell, M. Olette, N. Meysson, M. Chastent, A. Rst, P. Vicens: Rev. Met. **66**, 673 (1969)
6.15 H. Middendorf, J. Potschke, U. M. G. Frohberg: Metallwiss. Tech. **24**, 617 (1970)
6.16 A. D. Kulkarni, R. E. Johnson, G. W. Perbix: J. Inst. Metals **99**, 15 (1971)
6.17 W. L. Worrell: *Metal-Slag-Gas Reactions and Processes*, ed. by Z. A. Foroulis, W. W. Smeltzer (Electrochem. Soc. Inc., Princeton, N.J., 1975) pp. 822–833
6.18 R. Ruka, J. Weissbart: J. Electrochem. Soc. **109**, 723 (1962)
6.19 H. A. Liebhafsky, E. J. Cairns: *Fuel Cells and Fuel Batteries—A Guide to Their Research and Development* (John Wiley and Sons, New York 1968) p. 555
6.20 T. Takahashi: *Physics of Electrolytes*, Vol. 2, ed. by J. Hladik (Academic Press, London 1972) pp. 989–1049
6.21 E. F. Sverdrup, C. J. Warde, R. L. Eback: Ener. Conv. **13**, 129 (1973)
6.22 P. Kofstad: *Nonstoichiometry, Diffusion, and Electrical Conductivity in Binary Metal Oxides* (Wiley-Interscience, New York 1972)
6.23 P. G. Wahlbeck, P. W. Gilles: J. Am. Ceram. Soc. **49**, 180 (1966)
6.24 R. N. Blumenthal, J. Coburn, J. Baukus, W. M. Hirthe: J. Phys. Chem. Solids **27**, 643 (1966)
6.25 H. H. Möebius: Z. Chemie **4**, 81 (1964)
6.26 J. R. Johnson, C. E. Curtis: J. Am. Ceram. Soc. **37**, 611 (1954)
6.27 E. C. Subbarao, P. H. Sutter, J. Hrizo: J. Am. Ceram. Soc. **48**, 443 (1965)
6.28 R. C. Garvie: J. Am. Ceram. Soc. **51**, 553 (1968)
6.29 P. Duwez, F. H. Brown, F. Odell: J. Electrochem. Soc. **98**, 356 (1951)
6.30 C. Delamarre, M. Perez y Jorba: Rev. Hautes Temp. Refract. **2**, 314 (1965)
6.31 F. M. Spiridonov, L. N. Komissarova, A. G. Kocharov, V. I. Spitsyn: Zh. Neorg. Khim. **14**, 2535 (1969); Russ. J. Inorg. Chem. **14**, 1332 (1969)
6.32 R. Ruh, H. J. Garret: J. Am. Ceram. Soc. **50**, 257 (1967)
6.33 R. Roy, H. Miyabe, A. M. Diness: Bull. Am. Ceram. Soc. **43**, 255 (1964)
6.34 P. Duwez, F. Odell, F. H. Brown: J. Am. Ceram. Soc. **35**, 107 (1952)
6.35 C. F. Grain: J. Am. Ceram. Soc. **50**, 288 (1967)
6.36 H. A. Johansen, J. G. Cleary: J. Electrochem. Soc. **111**, 109 (1964)
6.37 C. Wagner, W. Schottky: Z. Phys. Chem. B**11**, 163 (1930)
6.38 R. A. Swalin: *Thermodynamics of Solids*, 2nd ed. (John Wiley and Sons, New York 1972)
6.39 F. A. Kröger: *The Chemistry of Imperfect Crystals*, Vols. 1–3, 2nd ed. (North-Holland Pub. Co., Amsterdam 1974)
6.40 H. Reiss: *Chemical and Mechanical Behavior of Inorganic Materials*, ed. by A. W. Searcy, D. V. Ragone, U. Colombo (Wiley-Interscience, New York 1970) pp. 493–521
6.41 F. A. Kröger, H. J. Vink: *Solid State Physics*, Vol. 3, ed. by F. Seitz, O. Turnbell (Academic Press, New York 1956) p. 307

6.42   B.C.H.Steele: *Solid State Chemistry*, Vol. 10 of Inorganic Chemistry Series, ed. by L.E.J.Roberts (Butterworths, London 1972) pp. 118–149

6.43   L.E.J.Roberts: *Nonstoichiometric Compounds*, Advan. Chem. Ser. 39 (Amer. Chem. Soc., Washington, D.C., 1963) pp. 66–73

6.44   H.H.Möebius: Z. Chem. **2**, 100 (1962)

6.45   D.M.Roy, R.Roy: J. Electrochem. Soc. **111**, 421 (1964)

6.46   F.Hund: Ber. Deutsch. Keram. Ges. **42**, 251 (1965)

6.47   J.Rudolph: Z. Naturforsch. A **14**, 727 (1959)

6.48   J.E.Bauerle: J. Chem. Phys. **45**, 4162 (1966)

6.49   M.F.Lasker, R.A.Rapp: Z. Phys. Chem. N. F. **49**, 198 (1966)

6.50   I.Bransky, N.M.Tallan: J. Am. Ceram. Soc. **53**, 625 (1970)

6.51   R.A.Rapp: *Thermodynamics of Nuclear Materials* (Inter. Atomic Energy Agency, Vienna 1968) pp. 559–585

6.52   B.C.H.Steele, C.B.Alcock: Trans. Met. Soc. AIME **223**, 1359 (1965)

6.53   F.Hund, W.Daerrwaechter: Z. Anorg. Allg. Chem. **265**, 67 (1951)

6.54   F.Hund, R.Mezger: Z. Phys. Chem. (Leipzig) **201**, 268 (1952)

6.55   J.M.Wimmer, L.R.Bidwell, N.M.Tallan: J. Am. Ceram. Soc. **50**, 198 (1967)

6.56   W.D.Kingery, J.Pappis, M.E.Doty, D.C.Hill: J. Am. Ceram. Soc. **42**, 393 (1959)

6.57   L.A.Simpson, R.E.Carter: J. Am. Ceram. Soc. **49**, 139 (1966)

6.58   W.H.Rhodes, R.E.Carter: J. Am. Ceram. Soc. **49**, 244 (1966)

6.59   H.H.Möebius, H.Witzmann, D.Gerlach: Z. Chem. **4**, 154 (1964)

6.60   R.E.Carter, W.L.Roth: *Electromotive Force Measurements in High-Temperature Systems*, ed. by C.B.Alcock (Institute of Mining and Metallurgy, London 1968) pp. 125–144

6.61   J.E.Bauerle, J.Hrizo: J. Phys. Chem. Solids **30**, 565 (1969)

6.62   J.E.Bauerle: J. Phys. Chem. Solids **30**, 2657 (1969)

6.63   T.Y.Tien, E.C.Subbarao: J. Chem. Phys. **39**, 1041 (1963)

6.64   E.C.Subbarao, P.H.Satter: J. Phys. Chem. Solids **25**, 148 (1964)

6.65   J.B.Wachtman, Jr., W.C.Corwin: J. Res. Nat. Bur. Std. **69A**, 457 (1965)

6.66   J.W.Patterson: J. Electrochem. Soc. **118**, 1033 (1971)

6.67   T.A.Ramanarayanan, W.L.Worrell: Can. Met. Quart. **13**, 325 (1974)

6.68   J.W.Patterson, E.C.Bogren, R.A.Rapp: J. Electrochem. Soc. **114**, 752 (1967)

6.69   C.Wagner: Z. Phys. Chem. B **21**, 25 (1933)

6.70   H.Schmalzried: *Thermodynamics*, Vol. 1 (Internat. Atomic Energy Agency, Vienna 1966) pp. 97–109

6.71   W.L.Worrell, J.L.Iskoe: *Fast Ion Transport in Solids, Solid Batteries and Devices*, ed. by W. van Gool (North-Holland Pub. Co., Amsterdam 1973) pp. 513–521

6.72   W.L.Worrell: to appear in *Proceed. of the Workshop on Electrocatalysis on Non-Metallic Surfaces* (Nat. Bur. Stand. Pub., 1976)

6.73   T.A.Ramanarayanan, W.L.Worrell: Met. Trans. **5**, 1773 (1974)

6.74   H.Schmalzried, A.D.Pelton: *Annual Review of Materials Science*, Vol. 2, ed. by R.A.Huggins (Annual Reviews, Inc., Palo Alto, Calif. 1972) pp. 143–180

6.75   B.C.H.Steele: *Electromotive Force Measurements in High-Temperature Systems*, ed. by C.B.Alcock (Institute of Mining and Metallurgy, London 1968) pp. 3–27

6.76   K.S.Goto, W.Pluschkell: *Physics of Electrolytes*, Vol. 2, ed. by J.Hladik (Academic Press, London 1972) pp. 539–622

6.77   W.L.Worrell: *Thermodynamics*, Vol. 1 (Internat. Atomic Energy Agency, Vienna 1966) pp. 131–143

6.78   S.C.Singhal, W.L.Worrell: *Metallurgical Chemistry*, ed. by O.Kubaschewski (Her Majesty's Stationary Office, London 1972) pp. 65–74

6.79   S.C.Singhal, W.L.Worrell: Met. Trans. **4**, 895 (1973)

6.80   S.C.Singhal, W.L.Worrell: Met. Trans. **4**, 1125 (1973)

6.81   H.Schmalzried: *Metallurgical Chemistry*, ed. by O.Kubaschewski (Her Majesty's Stationary Office, London 1972) pp. 39–64

6.82   C.Wagner, A.Werner: J. Electrochem. Soc. **110**, 326 (1963)

6.83   P.J.Meschter, W.L.Worrell: Met. Trans. **7**A, 299 (1976)

6.84   P.J.Meschter, W.L.Worrell: to appear in Met. Trans. (1977)

6.85   C.B.Alcock, T.N.Belford: Trans. Faraday Soc. **60**, 822 (1964)

6.86   T.N.Belford, C.B.Alcock: Trans. Faraday Soc. **61**, 443 (1965)

6.87   D.O.Raleigh: *Progress in Solid State Chemistry*, Vol. 3, ed. by H.Reiss (Pergamon Press, Oxford 1967) pp. 83–134

6.88   H.Rickert: *Electromotive Force Measurements in High-Temperature Systems*, ed. by C.B.Alcock (Institute of Mining and Metallurgy, London 1968) pp. 59–90

6.89   B.C.H.Steele: *Heterogeneous Kinetics at Elevated Temperatures*, ed. by G.R.Belton, W.L.Worrell (Plenum Pub. Corp., New York 1970) pp. 135–163

6.90   Y.D.Tretyakov, R.A.Rapp: Trans. Met. Soc. AIME **242**, 1235 (1969)

6.91   H.F.Rizzo, R.S.Gordon, J.B.Cutler: J. Electrochem. Soc. **116**, 267 (1969)

6.92   H.G.Sockel, H.Schmalzried: Ber. Bunsenges. Phys. Chem. **72**, 745 (1968)

6.93   F.E.Rizzo, J.V.Smith: J. Phys. Chem. **72**, 485 (1968)

6.94   H.F.Rizzo, R.S.Gordon, I.B.Cutler: *Mass Transport in Oxides*, ed. by J.B.Wachtman, A.D.Franklin (Nat. Bur. Stand. Spec. Pub. No. 296, 1968) pp. 129–142

6.95   K.Kiukkola: Acta Chem. Scand. **16**, 327 (1962)

6.96   T.L.Markin, R.J.Bones, V.J.Wheeler: Proc. Brit. Ceram. Soc. **8**, 51 (1967)

6.97   C.B.Alcock, S.Zador, B.C.H.Steele: Proc. Brit. Ceram. Soc. **8**, 231 (1967)

6.98   B.C.H.Steele: *Mass Transport in Oxides*, ed. by J.B.Wachtman, A.D.Franklin (Nat. Bur. Stand. Spec. Pub. No. 296, 1968) pp. 165–172

6.99   H.Schmalzried: Arch. Eisenhüttenw. **39**, 531 (1968)

6.100 H.Rickert, R.Steiner: Z. Phys. Chem. **49**, 127 (1966)

6.101 H.Rickert, A.A.El Miligy: Z. Metallk. **59**, 635 (1968)

6.102 H.Rickert, H.Wagner, R.Steiner: Chem. Ing. Tech. **38**, 618 (1966)

6.103 K.E.Oberg, M.Friedman, W.M.Boorstein, R.A.Rapp: Met. Trans. **4**, 61 (1973)

6.104 J.Osterwald, G.Schwarzlose: Z. Phys. Chem. N. F. **62**, 119 (1968)

6.105 S.Honna, N.Sano, Y.Matsushita: Met. Trans. **2**, 1494 (1971)

6.106 R.L.Pastorek, R.A.Rapp: Trans. Met. Soc. AIME **245**, 1711 (1969)

6.107 T.A.Ramanarayanan, R.A.Rapp: Met. Trans. **3**, 3239 (1972)

6.108 T.H.Etsell, S.N.Flengas: J. Electrochem. Soc. **118**, 1890 (1971)

6.109 F.A.Kröger: J. Electrochem. Soc. **120**, 75 (1973)

6.110 V.B.Tare, H.Schmalzried: Trans. Met. Soc. AIME **236**, 444 (1966)

6.111 N.Sano, S.Honna, Y.Matsushita: Met. Trans. **1**, 301 (1970)

6.112 J.Weissbart, R.Ruka: Rev. Sci. Instr. **32**, 593 (1961)

6.113 H.Schmalzried: Z. Electrochem. **66**, 572 (1962)

6.114 W.M.Hickam: Vacuum Microbalance Tech. **4**, 47 (1964)

6.115 H.H.Möebius: Z. Phys. Chem. (Leipzig) **230**, 395 (1965)

6.116 T.H.Etsel, S.N.Flengas: Met. Trans. **3**, 27 (1972)

6.117 D.Yuan, F.A.Kröger: J. Electrochem. Soc. **116**, 594 (1969)

6.118 K.E.Oberg, L.M.Friedman, W.M.Boorstein, R.A.Rapp: Met. Trans. **4**, 75 (1973)

6.119 E.Berkey, W.H.Reed, B.R.Grundy, M.H.Cooper: Trans. Am. Nucl. Soc. **12**, 609 (1969)

6.120 B.R.Grundy, E.Berkey, E.T.Weber, W.A.Ross: Trans. Am. Nucl. Soc. **14**, 186 (1971)

6.121 J.M.McKee, D.R.Vissers, P.A.Nelson, B.R.Grundy, E.Berkey, G.R.Taylor: Nucl. Tech. **21**, 217 (1974)

6.122 W.A.Fisher, W.Ackerman: Arch. Eisenhüttenw. **36**, 643 (1965);—**37**, 43 (1966)

6.123 K.Schwerdtfeger: Trans. Met. Soc. AIME **239**, 1276 (1967)

6.124 E.T.Turkdogan, R.J.Fruehan: Can. Met. Quart. **11**, 371 (1972)

6.125 W.A.Fisher, D.Janke: Arch. Eisenhüttenw. **41**, 1027 (1970)

6.126 W.Baukal: Chem. Ing. Tech. **41**, 791 (1969)

6.127 C.C.Sun, E.W.Hawk, E.F.Sverdrup: J. Electrochem. Soc. **119**, 1432 (1972)

7. Electrochemistry of Mixed Ionic-Electronic Conductors

L. Heyne

With 16 Figures

Various disciplines from the fields of physics and chemistry are required for a satisfactory description of the properties of mixed ionic-electronic conductors. A difficulty is that often the everyday concepts of the separate disciplines contain limiting conditions that are not fulfilled in the type of compounds discussed in this book. Consequently, before application of such concepts, their basic origins must be considered carefully, and from these it must be checked whether the application is justified or not.

It is the aim of the present chapter to do this; we shall study how principles from thermodynamics and from defect chemistry can be combined with those from semiconductor physics to build a "solid state electrochemistry" applicable to solids in which combined ionic and electronic transport can take place. This is feasible because the conductivity mechanism imposes some restrictions on the otherwise wide variety of possible combinations of defect chemical parameters. In this way a number of conclusions of general validity can be made for this group of compounds.

There are several ways by which solids where ionic conductivity plays an important part can be divided into classes. One possible classification is given in Table 7.1. This scheme serves to define some terms that will be used throughout this chapter. In contrast to the usual situation in solution, ionic conduction in solids is accompanied by an electronic counterpart; only if the ionic contribution to the total electrical conductivity is the major part shall we speak about a "solid electrolyte". Other cases will be referred to as "mixed conductors".

Roughly speaking, the reasons for ionic and for electronic conduction are independent. The first is mainly related to crystal structure; the latter is determined by the electronic bandgap, which depends more on the individual properties of the constituent ions. Thus if a crystal structure is favorable for fast ionic movement, the high ionic mobility in it may as well be accompanied by a very low or by a very high electronic conductivity. That is, we find series of compounds with comparable ionic conductivity, in which the electronic conductivity varies in a series from very low (the compound is an electrolyte in that case) to quasi-metallic. We will pay attention to both extremes. The first kind may find application as the electrolyte in galvanic cells; the other might form a favorable class of materials for an electrode in such cells.

Both groups are subdivided further into two classes. In the first the ionic movement is only by point defects in the crystal lattice. These lattice defects may for example be vacant lattice positions, or ions placed at normally unoccupied

Table 7.1. Solid ionic conductors

Point defect type		Cation-disordered sublattice type	
Dilute	Concentrated	Low temperature	High temperature
1) Electrolytes			
NaCl	$ZrO_2 \cdot CaO$	$RbAg_4I_5$	Li_2SO_4
AgBr	CaF_2	$\beta\text{-}Al_2O_3$	
LaF_3			
2) Mixed conductors			
$\beta\text{-}Ag_2S$	UO_2	$\alpha\text{-}Ag_2S$	

positions, so-called interstitial ions. They form only a small fraction of the normally placed ions and their concentration can be modified by doping with foreign ions. The defects can move by a jump mechanism often involving neighboring lattice ions as well. However, most of the ions at regular lattice positions cannot contribute to conduction or diffusion processes.

In the other class there exists a large excess of available lattice positions for the mobile kind of ion. The latter occupy these in an unordered way, and moreover, can change position very easily. A distinction between normal and interstitial positions in such a sublattice is meaningless from a defect chemical point of view. All ions of the type concerned contribute to the electrical conduction and diffusion processes. The magnitudes of the ionic conductivity and self-diffusion coefficient are comparable to those of liquid electrolytes. In such cases we will speak about "cation-disordered sublattice conductors".

The two classes can be subdivided further as shown in the table, the cation-disordered sublattice compounds according to the temperature region where high conductivity exists, the defect type compounds according to the concentration of lattice defects that can be introduced. A temperature is considered high if it is near the melting point of the substance. Concentrated defect compounds resemble cation-disordered sublattice conductors in the respect that they show high lattice disorder, and that strong interaction between the moving ions is present.

Typical examples of the four groups are given at the bottom of the table: in row 1) electrolytes, in row 2) mixed conductors.

7.1 Thermodynamics of Electronic and Ionic Charge Carriers

In solid electrolytes and related compounds the concentrations of ionic and electronic charge carriers are determined by the chemical composition of the substance, as is the case with electrolytic solutions and with semiconductors. An important aspect of compounds with high ionic mobility is the rapidity with which they can respond, at relatively low temperatures, to externally imposed conditions. Thermodynamic equilibrium, for example with an ambient atmosphere, is often established in practice. Equilibrium thermodynamics therefore is of great practical importance, and can be used to derive quantitative relations between carrier concentrations and chemical potentials of components.

A consequence is that the composition is liable to changes, and that significant values for electronic conductivity parameters can be determined or specified only if care is taken to control the sample composition or environmental conditions in one way or another. In other words: *physical properties of mixed conductors are defined only if the chemo-thermodynamic conditions are well controlled.* This is in contrast with, for instance, semiconductor materials in which, at the temperature of their use, ions do not move under applied fields, and the chemical equilibria are completely frozen in. Only at higher temperatures, for example those used for the fabrication of semiconductor devices, in the sintering of ceramic products, the firing of luminescent powders, etc., do all common compounds become mixed conductors to some extent. Consequently, many of the considerations given in this chapter will apply to such cases also.

In what follows, we will use chemical potentials and activities both of neutral components of the system under consideration, and of charged species. To stress the difference and to avoid confusion, the quantities referring to neutral components will always be marked with an asterisk *. They are real thermodynamic quantities in the sense that these neutral components can really be added as such to a system; this in contrast with charged constituents like ions or charged defects which must be added in pairs of opposite charge to maintain electroneutrality (see for example, *Guggenheim* [7.1]). Chemical potentials and activities without the asterisk will always refer to charged particles. Although they are thermodynamically undefined for the above-mentioned reason, they are of great value for considerations based on a physical model.

Because we will think mostly in terms of single particles such as vacancies, interstitials, electrons and (electronic) holes, we choose the atomic unit instead of the gram mol as the mass unit to define concentrations, chemical potentials, etc. In this way we also obtain equations that conform to those used in semiconductor physics.

This has the further consequence that in the formulas, the gas constant R is replaced by Boltzmann's constant k and the Faraday constant by the absolute value of the electronic charge e.

The ionic quantities can be those for the conventional ions. But they can also refer to lattice defects with an effective charge, which in the case of crystalline solids often take over the role of the conventional ions in thermodynamic relations, see *Kröger* [Ref. 7.2, Sect. 21.2]. What is meant will always be indicated by an appropriate index.

A complete list of the symbols used is given at the end of this chapter.

7.1.1 Carrier Concentrations

We shall now apply general thermodynamic concepts to calculate charge carrier concentrations and, through multiplication by their charge and mobility, also partial conductivities. The concentration of electronic charge carriers is influenced by external conditions such as partial pressure of a component; under certain circumstances this may lead to such an increase in electronic conductivity (either n-type or p-type) that an electrolyte changes into a mixed conductor.

It will be our main concern to determine the boundaries in temperature and compositional parameters where this occurs, and thus where the "electrolytic domain" ends. To keep the considerations definite and simple, a binary compound MX with Frenkel disorder will be taken as an example. It will be assumed that M and X are monovalent, and that also the ionic defects have unity charge. This means that the dominant defects are interstitial M^+-ions, in accordance with the notation of *Kröger* and *Vink* [7.2–4] denoted by M_i^+, and M^+-ion vacancies denoted by V_M^-. These assumptions are made for the sake of a simple presentation only. Along completely analogous lines, equivalent results can be obtained if other assumptions about the type of disorder were used as a starting point.

Furthermore it should be recognized that disordered sublattice electrolytes can be considered as compounds with an extremely high degree of Frenkel disorder. Therefore, the concepts "vacancy" and "interstitial" may be retained in this general consideration, and, although the assignments "regular lattice site" (which can be vacant) and "interstitial lattice site" are arbitrary in such compounds, the following discussion remains applicable.

We assume that the following three independent defect chemical equilibrium reactions can occur:

$$M_i^+ + V_M^- \rightleftharpoons M_M + V_i \qquad (7.1)$$

$$e^- + h^+ \rightleftharpoons 0 \qquad (7.2)$$

$$M_i^+ + e^- \rightleftharpoons M_i \rightleftharpoons M_{external} + V_i. \qquad (7.3)$$

The first represents the generation-annihilation reaction of metal interstitials and vacancies as postulated for Frenkel disorder. Annihilation of a Frenkel pair [left side of (7.1)] gives rise to the generation of an occupied M-lattice position and an empty interstitial site, the normal lattice condition. Equation (7.2) represents the electron-hole equilibrium. Equation (7.3) states how interstitial ions can exchange with a neutral environment; the charge balance is retained by a combination with an electron.

The condition of thermodynamic equilibrium can then be expressed in terms of chemical potentials as follows[1].

$$\mu_{M_i^+} + \mu_{V_M^-} = \text{constant} \qquad (7.4a)$$

$$\mu_{e^-} + \mu_{h^+} = 0 \qquad (7.4b)$$

$$\mu_{M_i^+} + \mu_{e^-} = \mu_M^*. \qquad (7.4c)$$

---

[1]   The use of chemical potentials instead of electrochemical potentials is justified because here equivalent numbers of oppositely charged particles in the same phase are involved. The electrostatic terms in the electrochemical potentials thus cancel.

If desired, μ_X^* can be introduced instead of μ_M^*, with corresponding adaption of sign and zero level [see (7.20)].

If the defect concentrations are low, the chemical potentials can be expressed in terms of concentrations c_i by the "ideal" law

$$\mu_i = \mu_i^0 + kT \ln c_i. \tag{7.5}$$

Here the index $_i$ indicates the defect species and μ_i^0 is a standard potential. If referring to electrons, the equation indicates validity of Boltzmann's statistics. Considered from the electronic point of view the material is then a normal "nondegenerate" semiconductor (see, e.g., [7.5–8]). Because of the low electronic concentration, electrolytes always fall in this category. With ideal behavior the set of (7.4) can be replaced by the following "mass action" formulas:

$$[M_i^+][V_M^-] = v_0^2 \tag{7.6a}$$

$$np = n_0^2 \tag{7.6b}$$

$$n[M_i^+] = a' \exp - (\mu_X^*/kT) = aP_X^{-1}. \tag{7.6c}$$

Here v_0^2, n_0^2, a, and a' are concentration independent quantities.

We consider the partial pressure P_X (or the chemical potential μ_X^*) as the independent variable in these three equations. A fourth equation, needed to determine the four concentrations involved, follows from the requirement of electroneutrality. Because the crystal as a whole is neutral we may write this condition as

$$n + [A_X^-] + [V_M^-] = p + [M_i^+]. \tag{7.7}$$

Here n and p are the concentrations of "free" electrons and holes, respectively. Furthermore, it is assumed that a certain number of divalent foreign negative ions A^{2-} is present. Replacement of a regular X^- ion by such a foreign one and capture of an electron results in an effectively negatively charged "acceptor center" A_X^-. In compounds with ions of higher valency, other types of acceptor defects may occur (e.g., Ca_{Zr}^{2-} in ZrO_2); in pure compounds $[A_X^-] = 0$.

If the relations between concentrations and chemical potentials or the material constants in (7.6) are known, the four concentrations $[M_i^+]$, $[V_M^-]$, n and p can be calculated from the four equations, (7.7) with either the set (7.4) or (7.6). Many examples of such calculations for actual compounds, often involving more types of defects, can be found in Kröger's book on defect chemistry [7.2].

7.1.2 Disorder and Conductivity Types

In practice the solution of the system of equations is relatively simple because in (7.7) only one term on either side dominates, and this irrespective of whether the general case (7.4) is considered, or ideal behavior (7.6) is assumed. Consequently, six limiting forms of (7.7) can be considered, in each of which all but one term on

Table 7.2. Classification of disorder and conductivity types

Limiting neutrality equation	Disorder type	Conductivity type
(7.8) $p=n$	Intrinsic electronic	Intrinsic semiconductor
(7.9) $p=[A_X^-]$	Acceptor dominated electronic	p-type semiconductor
(7.10) $n=[M_i^+]$	n-type mixed ionic electronic	n-type semiconductor
(7.11) $p=[V_M^-]$	p-type mixed ionic electronic	p-type semiconductor
(7.12) $[M_i^+]=[V_M^-]$	Intrinsic ionic	Ionic conductor[a] or semiconductor
(7.13) $[M_i^+]=[A_X^-]$	Acceptor dominated ionic	Ionic conductor or semiconductor

[a]   According to the definitions given in Table 7.1 the ionic conductor may be an electrolyte or a mixed conductor.

each side are neglected [7.9]. The six limiting forms characterize six main types of disorder as shown in Table 7.2.

In electrolytes the highly mobile species e^- and h^+ cannot dominate, so only the last two types of disorder remain possible. The last one in the table, in which the defect concentration is fixed by the amount of doping, is typical for concentrated defect electrolytes (see Table 7.1) such as stablized zirconia. The previous one, (7.12), with equal concentrations of corresponding defects is typical for pure defect electrolytes, but also for cation-disordered sublattice electrolytes. In the latter case the defect concentrations are constant and determined by the structural disorder. For pure defect electrolytes they are constant because of the simultaneous validity of (7.6a) and (7.12). Note that in this respect "pure" means that the constant v_0 which determines the maximum value of $[V_M^-]$ is larger than the impurity (dopant) concentration $[A_X^-]$.

Thus in all cases where binary compounds show electrolytic or mixed conduction $\mu_{M_i^+}$ *and* $\mu_{V_M^-}$ *are constants with respect to variation of the independent variable* μ_X^*. In our model of the defect crystal lattice, the defects represent the only mechanism by which energetic and statistical properties of the constituent ions may vary. Thus we arrive at the important conclusion that in ionic conductors also the *chemical potentials of the ions must be independent of compositional variations* induced by changes in component chemical potential. Summarizing we have

$$\Delta\mu_{M_i^+}=0 \quad \Delta\mu_{M^+}=0$$
$$\Delta\mu_{M_V^-}=0 \quad \Delta\mu_{X^-}=0. \tag{7.14}$$

Pursuing the description in terms of *ionic* chemical potentials, we consider the more general case of the ionization equilibrium of a z-valent cation M^{z+}

$$M\rightleftharpoons M^{z+}+ze^-. \tag{7.15}$$

In terms of ionic chemical potentials the equilibrium condition reads

$$\mu_M^*=\mu_{M^{z+}}+z\mu_e, \tag{7.16}$$

or in terms of variations

$$\Delta\mu_M^* = \Delta\mu_{M^{z+}} + z\,\Delta\mu_e. \tag{7.17}$$

Note that (7.16) is the ionic equivalent of the defect equation (7.4c).

The quantity $\mu_{M^{z+}}$ is the kind of chemical potential which is almost exclusively considered in the field of classical electrochemistry. There it can be varied by altering the concentration of the electrolytic solution. In that field μ_e plays a role only in typical redox systems, and μ_M^* is hardly ever considered. In contrast, we saw in (7.14) that in solid ionic conductors the chemical potential of the ions is fixed, and that μ_M^* is the quantity that can be controlled experimentally, e.g., through μ_X or the partial pressure of a component, and therefore deserves most attention.

Because of the constancy of $\mu_{M^{z+}}$, variations in the chemical potential of the neutral component M and that of the electrons are directly coupled:

$$\Delta\mu_e = 1/z\,\Delta\mu_M^*. \tag{7.18}$$

Of course, equivalent reasoning can be used to reach similar expressions involving μ_X^* and/or μ_h. Note that μ_e is equivalent with the distance of the Fermi level to the edge of the conduction band in a semiconductor.

In electrolytes, with their low electronic concentrations, the electrons exhibit ideal behavior (Boltzmann's statistics), so that μ_e varies linearly with log n. Since μ_M^* shows the same dependence on P_M (metal partial pressure), we conclude

$$n \sim (P_M)^{1/z}. \tag{7.19}$$

For a binary compound $M_y X_z$ application of Gibbs-Duhem's equation [Ref. 7.1, p. 25]

$$y\,\Delta\mu_M^* + z\,\Delta\mu_X^* = 0 \tag{7.20}$$

yields the equivalent expression

$$n \sim (P_X)^{-1/y}. \tag{7.21}$$

For constant mobility the electronic conductivity follows the same type of law.

We stress that for the derivation of these important dependences, we did not use any detailed model for the solid. The only restriction introduced was in the transition from (7.1) to (7.4a), where the concentrations of normal lattice ions and normal interstitial positions were considered as constants. This means that the changes in these concentrations due to composition variation should be negligible. This is true for compounds with a narrow homogeneity range. Electrolytes and electronically nondegenerate mixed conductors always belong to this category. The argument breaks down for compounds with variable stoichiometry such as tungsten bronzes or wüstite ($Fe_{1-x}O$), which have metallic conductivity.

7.1.3 Graphical Representations, Conductivity Domain Boundaries

We now consider defect-type conductors more closely by solving the set (7.6) with each of the limiting forms of the neutrality equation given in Table 7.2. Because dominating ionic conductivity can occur only if the minimum electronic concentrations (being met when $n = p = n_0$) are much lower than the ionic defect concentrations, the constants in (7.6a and b) must satisfy: $n_0 \ll v_0$. In what follows we will always start from this assumption. The consequence is that intrinsic electronic disorder (first in Table 7.2) is ruled out. In Table 7.3, P_X dependences obtained for electron and hole concentrations are collected for the other five types of disorder. Note that for the first one in this table the electron concentration is decreasing with increasing X partial pressure, so that there comes a point where the limiting electroneutrality equation (7.10) is no longer a good approximation to (7.7), and either (7.12) or (7.13) takes over, depending on whether $v_0 > [A_X^-]$ or not (pure or doped compound). This means that a new type of disorder starts to dominate. In a similar way, a further increase in partial pressure leads to new transitions in disorder type.

Such transitions, together with the concentration dependences of the various defects, are clearly illustrated in the graphical representations of Figs. 7.1a (pure compound) and 7.2a (doped compound); so-called *Brouwer* diagrams [7.9]. The transitions are represented by kinks in the concentration curves. In reality such kinks should be replaced by smooth joints of the straight line segments (which could be calculated by retaining one more term in the neutrality equation). However, since the logarithmic scales extend over many decades, the figure gives a fairly good representation of the real situation.

The partial conductivities plotted in Figs. 7.1b and 7.2b are obtained by multiplication of the concentrations of Figs. 7.1a and 7.2a with the respective charges and mobilities. Since the latter are much higher for electronic than for ionic charge carriers, the main effect is that in the logarithmic σ-representations, the electronic curves are shifted upwards with respect to the ionic ones. For the sake of simplicity pairwise equal mobilities are assumed for electrons and holes and for interstitials and vacancies.

The various ranges of defect type as well as of conductivity type indicated in Table 7.2 are clearly evident from these graphs. We see that the electrolytic conductivity domain is appreciably narrower than the range of ionic disorder, and furthermore, that mixed conductor regions separate the electrolytic domain from pure semiconductor regions. If the ionic defect mobilities were very low, as is the case with most "normal" compounds, the separation of the electronic and ionic curves would be much wider and the midregion of the diagram would correspond to "insulators".

In principle, the formulas used, and consequently also the graphs based on them, are valid only for compounds with defect concentrations low enough to allow application of mass-action equations [ideal thermodynamic laws, (7.6)]. They are bound to fail for compounds with a high concentration of defects or with a cation-disordered sublattice. However, the parts of the graphs represent-

Table 7.3. Partial pressure dependences of electronic carriers

Limiting neutrality equation	Disorder type	P_X range	Electron concentration n	Hole concentration p
(7.10) $n=[M_i^+]$	n-type mixed ionic electronic	Low	$v_0 n_0 (aP_X)^{-1/2}$	$n_0/v_0 (aP_X)^{1/2}$
(7.12) $[M_i^+]=[V_M^-]$ $=v_0$	Intrinsic ionic	Medium	$v_0 n_0^2 (aP_X)^{-1}$	$1/v_0 (aP_X)$
(7.13) $[M_i^+]=[A_X^-]$	Acceptor dominated ionic	Medium	$v_0^2 n_0^2 (aP_X)^{-1}/[A_X^-]$	$1/v_0^2 (aP_X)[A_X^-]$
(7.9) $p=[A_X^-]$	Acceptor dominated electronic	High	$n_0^2/[A_X^-]=$ constant	$[A_X^-]=$ constant
(7.11) $p=[V_M^-]$	p-type mixed ionic electronic	High	$n_0^2 (aP_X)^{-1/2}$	$(aP_X)^{1/2}$

The third and fourth rows apply only if the dope concentration $[A_X^-]$ exceeds the intrinsic ionic defect concentration: $[A_X^-] > v_0$.

ing solid electrolyte or mixed conductor behavior remain applicable. This is so because, first, these parts still belong to the region of dominating ionic disorder so that (7.12) or (7.13), but also (7.14), remain valid, notwithstanding the nonideal thermodynamics of the defects. Secondly, the electronic carriers have such low concentrations that ideal laws can be used for them. So the real electron and hole concentration dependences are still well represented by the curves given, and this is in full accordance with (7.19) and (7.21) discussed earlier.

Only if a transition to a mixed electronic-ionic disorder situation takes place in a cation-disordered sublattice electrolyte would the situation change drastically. Metallic conduction results because of the high concentrations involved; the electronic system is degenerated and, although still $n=[M_i^+]$, the nonideality yields curvature of the n vs μ_X^* dependence and even more strongly deviating conductivity curves. Such cases will be discussed in Section 7.1.5.

Electronic Transport Numbers of Electrolytes

Transport numbers, in logarithmic measure, can be read from the conductivity diagrams as the distance between the ionic and electronic curves. The electronic transport number t_e is minimum at the point of intrinsic electronic conductivity where the n- and p-type lines cross. In equating the values for n and p of the second row of Table 7.3, we find that this point occurs at

$$P_0 = n_0 v_0/a, \quad \text{if} \quad [A_X^-] < v_0 \tag{7.22}$$

for a pure compound. Doing the same for an acceptor doped electrolyte (third row) we find for the corresponding partial pressure of component X

$$P_0' = P_0 v_0/[A_X^-], \quad \text{if} \quad [A_X^-] > v_0. \tag{7.23}$$

Thus acceptor doping shifts the pressure of optimum electrolytic behavior (t_e minimum) towards lower P_X inversely proportional to the dopant concentration.

a.

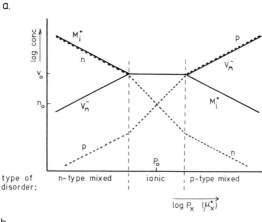

b.

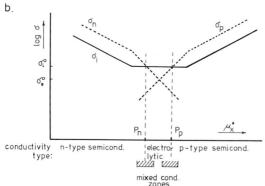

Fig. 7.1. (a) Defect diagram of pure compound MX with Frenkel disorder. Dotted lines represent electronic defects, solid lines ionic defects. (b) Corresponding conductivity diagram. Electrolytic domain boundaries are within range of ionic disorder

This minimum electronic transport number improves (i.e., drops), because, as a rule, the ionic conductivity increases with doping, while the intrinsic electronic conductivity is of course unaffected.

At a constant pressure the value P_X with respect to P'_0 is important. We first consider cases for which $P_X > P'_0$. Then the electronic conductivity is p-type and the shift in P'_0 as a consequence of dopant variation (7.23) leads to a proportional change in the hole conductivity. This is so for dilute defect electrolytes. With high defect concentration the strict proportionality in the shift of the intrinsic point is no longer there [because M_i^+ becomes nonideal, and its concentration may not be retained in (7.6c)]. However, the trend remains, so that the hole conductivity increases with increasing acceptor doping. Consequently, effects that are proportional to σ_p, such as the oxygen permeation through stabilized zirconia, must also increase with higher acceptor doping (Ca or Y in ZrO_2) [7.10].

For ideal behavior of defects, both the ionic defect and the hole concentrations vary proportionally with the acceptor concentration, and thus the electronic transport number remains constant at pressures in this range. But again this argument breaks down for higher defect concentrations. Apart from the nonlinear shift in P'_0, there is the decreasing mobility of the ionic

a.

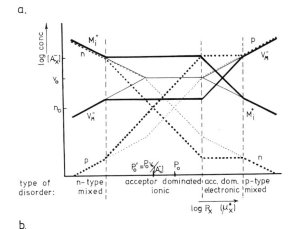

type of disorder: n-type mixed | acceptor dominated ionic | acc. dom. electronic | p-type mixed

$\log P_X \; (\mu_X^*)$

b.

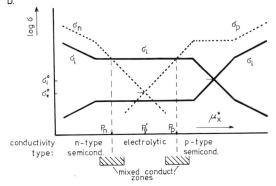

conductivity type: n-type semicond. | electrolytic | p-type semicond.

mixed conduct. zones

Fig. 7.2. (a) Defect diagram of acceptor doped compound MX with Frenkel disorder. Lightly drawn lines show case of pure compound for comparison. (b) Corresponding conductivity diagram. The electrolytic domain is extended to lower μ_X^* as a result of doping

carriers due to the stronger mutual interaction. Both effects tend to cancel, but for stabilized zirconia, for example, the mobility drops so fast with increasing stabilizer concentration [7.11] that we expect the electronic transport number to increase under these circumstances.

At partial pressures such that $P_X < P_0'$ (n-type electronic conductivity) the shift in P_0' results in changes in n and σ_n opposite to those discussed above for electron holes. Increase in the acceptor dopant concentration lowers the electron concentration, and, because the mobile defect concentration is enhanced at the same time, the transport number improves strongly. A high stabilizer content seems to be favorable for the application of zirconia at low oxygen partial pressures.

Domain Boundaries

The electrolytic domain boundaries are very important for the judgment of the practical usability of an electrolyte under certain environmental conditions. Their positions on the partial pressure scale P_n and P_p (see Figs. 7.1 and 2) can be found quite simply from the equations given in Table 7.3 (second row). Using the quantities σ_e^0 and σ_i^0 for the intrinsic electronic and ionic conductivities, respectively (which occur at the partial pressure P_0 in the pure compound, as also indicated in Fig. 7.1), we see that in a pure compound

$$\sigma_n = \sigma_e^0 (P_0/P_X), \tag{7.24}$$

$$\sigma_p = \sigma_e^0 (P_X/P_0). \tag{7.25}$$

Equating these conductivities with the ionic conductivity $\sigma_i^0 = \sigma_i$, we obtain for the domain boundaries in an undoped compound

$$P_n = (\sigma_e^0/\sigma_i) P_0 \quad \text{and} \quad P_p = (\sigma_i/\sigma_e^0) P_0. \tag{7.26}$$

The logarithmic width of the electrolytic domain thus becomes

$$P_p/P_n = (\sigma_i/\sigma_e^0)^2. \tag{7.27}$$

For a doped compound the n and p dependences from the third row of Table 7.3 lead to

$$\sigma_n = \sigma_e^0 (P_0/P_X)(v_0/[A_X^-]) = \sigma_e^0 (P_0'/P_X) \tag{7.28}$$

$$\sigma_p = \sigma_e^0 (P_X/P_0)([A_X^-]/v_0) = \sigma_e^0 (P_X/P_0') \tag{7.29}$$

when (7.23) is also introduced. At the same time the ionic conductivity has increased with respect to that of the pure compound

$$\sigma_i = \sigma_i^0 ([A_X^-]/v_0). \tag{7.30}$$

Equating the electronic conductivities with this value yields for the domain boundaries in the acceptor dominated electrolyte

$$P_n = (\sigma_e^0/\sigma_i^0)(v_0^2/[A_X^-]^2) P_0 \quad \text{and}$$

$$P_p = (\sigma_i^0/\sigma_e^0) P_0. \tag{7.31}$$

From comparison with (7.26) we see that the high pressure domain boundary is not influenced by acceptor doping. However, the low pressure boundary is shifted to lower pressures in an inverse quadratic dependence of dopant concentration. Consequently, the domain width, which is given by

$$P_p/P_n = (\sigma_i^0/\sigma_e^0)^2 ([A_X^-]^2/v_0^2) \tag{7.32}$$

is increasing quadratically with dopant concentration.

When we introduce the actual ionic conductivity σ_i instead of that of the undoped compound via (7.30) we obtain

$$P_p/P_n = (\sigma_i/\sigma_e^0)^2,$$
(7.33)

the same expression as for the undoped compound.

It will be clear that the shifts take the opposite direction when doping is carried out with donor-type impurities. Furthermore, we note that again nonideal behavior of the defects modifies the functional dependence, but the trends remain the same in concentrated defect electrolytes. Because doping parameters enter the limiting neutrality equation only if the dopant concentration exceeds v_0, we expect only minor influence of impurities in cation-disordered sublattice electrolytes.

The expressions (7.24–33) are valid only for the monovalent species with ideal thermodynamic behavior, for which the Brouwer diagrams of Figs. 7.1 and 7.2 were constructed. When we refrain from expressing the dependence on dopant concentration, the equations can be put on a more general foundation. We start from equations expressing the electron concentration as a function of partial pressure derived before for a compound M_yX_z [(7.21)] and its equivalent for holes $p \sim (P_X)^{1/y}$, which are valid for any (binary or quasi-binary) electrolyte. Following the same lines as used for the derivation of (7.33) we obtain

$$\sigma_n = \sigma_e^0 (P_0'/P_X)^{1/y}$$
(7.28a)

$$\sigma_p = \sigma_e^0 (P_X/P_0')^{1/y}$$
(7.29a)

$$\left. \begin{aligned} P_n &= (\sigma_e^0/\sigma_i)^y P_0' \\ P_p &= (\sigma_i/\sigma_e^0)^y P_0' \end{aligned} \right\}$$
(7.31a)

$$P_p/P_n = (\sigma_i/\sigma_e^0)^{2y}.$$
(7.33a)

These equations are valid for solid electrolytes with a high defect concentration as well as for those with a cation-disordered sublattice.

Three-Dimensional Representations

When curves such as those of Figs. 7.1b or 7.2b are constructed for different temperatures, where the mass constants and the mobilities are modified, a three-dimensional graph, first introduced by *Patterson* [7.12], can be obtained. The ionic, the n-type and the p-type conductivities appear as planes, see Fig. 7.3. The boundaries between the electrolytic and electronic conductivity domains are represented by the intersections of these planes. Their positions are also clear from the projection onto the base plane $(1/T - \mu_X^*$ plane) as shown in Fig. 7.4. The two plots cover only the region of ionic disorder, but apply to both a pure compound and to one with a constant dopant concentration.

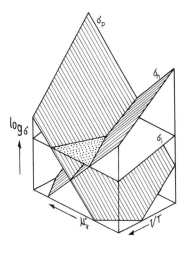

Fig. 7.3. Conductivity diagram after *Patterson* [7.12]. Domain boundaries are given as intersections of the ionic and electronic conductivity planes (from *Heyne* [7.13])

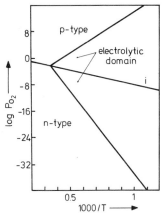

Fig. 7.4. Domain boundaries of stabilized zirconia as a projection on the $\mu_X^* - 1/T$ plane. i = line of intrinsic electronic conductivity (from *Heyne* [7.13])

Instead of the three-dimensional $\log \sigma$ vs μ_X^* and $1/T$ plot, we can also construct an isothermal $\log \sigma$ vs μ_X^* and $\log$ dopant concentration $[A_X^-]$ dependence. Its approximate shape is given in Fig. 7.5. We recognize an ionic conductivity plane consisting of two parts. One is horizontal which means that the ionic conductivity is independent of both μ_X^* and $[A_X^-]$; the dopant concentration $[A_X^-]$ is smaller than v_0 in this region and (7.12) is valid. The second part shows the upward slope reflecting the increase with dopant concentration when $[A_X^-] > v_0$ but σ_i remains independent of μ_X^*; (7.13) holds. As in the preceding graph, domain boundaries are lines of intersection of the ionic conductivity plane with the electronic ones. In both plots the intersection of the n- and p-type planes represents situations where the electronic conductivity is intrinsic. The corresponding conductivity value is independent of μ_X^* as well as of dopant concentration. Its value is determined mainly by the electronic bandgap. The oblique position in the horizontal plane in Fig. 7.5 above the critical

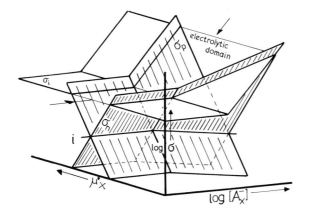

Fig. 7.5. Isothermal conductivities as a function of μ_X^* and $[A_x^-]$. At $[A_x^-] = v_0$ all conductivity planes bend

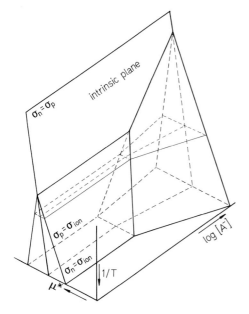

Fig. 7.6. Domain boundary planes in $\mu_X^* - [A_x^-] - 1/T$ space. The ionic domain narrows at higher temperatures because of the high activation energy of the electronic conductivity

concentration v_0 reflects the combined influence of dopant and partial pressure discussed earlier.

A clearer insight into the dependence of the domain boundaries on the three independent parameters can be obtained when we plot the boundaries (lines in the two preceding three-dimensional graphs) in μ_X^*, $1/T$, $[A_X^-]$ space. Such a plot can be constructed by projecting the domain boundaries of Fig. 7.5 onto the base plane. When we do this for various temperatures and erect a (reciprocal) T-axis perpendicular to the base plane, we obtain the plot of Fig. 7.6. The two planes $\sigma_p = \sigma_i$ and $\sigma_n = \sigma_i$ separate the tentlike space of the electrolytic domain from the electronic region above. The rising part represents the increase in defect concentration and ionic conductivity upon increased doping. The electrolytic conductivity domain widens substantially in this region.

The high temperature limit of the electrolytic domain evident in the last plot exists because of the higher activation energy of the electronic conductivity as compared with the ionic one. The high activation energy of electronic conduction results from the high bandgap required for the low electronic conductivity of a good electrolyte. The low energy of activation of the ionic conductivity corresponds to the easy way in which high mobility ions in good solid electrolytes can penetrate intersite barriers in the crystal.

7.1.4 Requirements for Low Electronic Transport Number

From the three-dimensional plots we see that electrolytic domains exist only because the line representing intrinsic electronic conductivity may be situated below the ionic conductivity planes. Because the position of this intrinsic line is determined almost exclusively by the value of the electronic bandgap, a minimum requirement for this bandgap can be formulated for solid electrolytes.

A semiquantitative evaluation starts from the expression [7.5–8]

$$n_0^2 = N_c N_v \exp(-eB/kT), \tag{7.34}$$

where n_0^2 is the same constant as in (7.6b). B is the width of the forbidden zone between the bands, the so-called bandgap. N_c and N_v are the effective densities of electronic states in the conduction and valence bands, respectively, and can be calculated from

$$N_{c,v} = 2(2\pi m_{c,v} kT/h^2)^{3/2} = 4.83 \cdot 10^{15} T^{3/2} (m_c/m)^{3/2}, \tag{7.35}$$

where h is Planck's constant, m the electronic mass and m_c and m_v are the effective masses of electrons and holes.

Assuming equal mobility u_e for electrons and holes again, we obtain for the intrinsic conductivity

$$\sigma_e^0 = 2n_0 e u_e. \tag{7.36}$$

Now we define, more or less arbitrarily, a "good" electrolyte as a compound having an ionic conductivity of about 0.01 $[\Omega\,\mathrm{cm}]^{-1}$ and an electronic transport number below 10^{-4} at the temperature concerned. In that case the electronic conductivity must be lower than 10^{-6} $[\Omega\,\mathrm{cm}]^{-1}$. We introduce a mobility estimate of 1 cm^2/Vs, a low value, which corresponds to an electron mean free path of roughly one interatomic distance as may be expected in the highly disordered crystal lattices characteristic for good ionic conductors. Furthermore, taking the value of the mass of a free electron for m_c and m_v, (7.34–36) then yield the following quantitative estimate for the minimum allowable value for the bandgap of a good electrolyte:

$$\begin{aligned} B_{0,\min} &= 2kT/e \ln(1.548 \cdot 10^{-3} T^{3/2}/\sigma_{e,\max}) + 5 \cdot 10^{-4}(T-273) \\ &= 1.72 \cdot 10^{-4} T \ln(1548 T^{3/2}) + 5 \cdot 10^{-4}(T-273). \end{aligned} \tag{7.37}$$

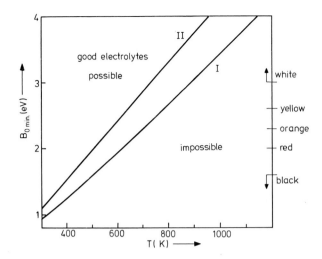

Fig. 7.7. Minimum value of bandgap for good electrolytes as a function of temperature (Curve I). A safety margin of a factor 100 is used for Curve II (from *Heyne* [7.13])

The last term originates from a temperature dependence of the bandgap, which was assumed to be of the form

$$B = B_0 - 5 \cdot 10^{-4}(T - 273), \tag{7.38}$$

where B_0 represents the bandgap at $0\,°C$. Curve I in Fig. 7.7 shows how the minimum allowable bandgap calculated according to (7.37) varies with temperature. Note that, because of the connection between the electronic bandgap and the color of a compound (the optical absorption edge), an approximate color scale could also be indicated in the figure.

The curve divides the coordinate field into two regions in which "good" electrolytes are possible or not. It appears that the position of the boundary line depends only weakly on the quantitative assumptions made for its calculation because of the dominating influence of the exponential dependence on B, (7.34). Furthermore, we see that the curve is nearly straight. *Therefore, it appears that the electronic bandgap of "good solid electrolytes" is always larger than $T/300$ eV* [7.13].

Of course, this is a necessary but not a sufficient requirement. We saw that an unfavorable environment may shift the value of P_X from its optimum value P_0 at the $n-p$ boundary to perhaps outside the electrolytic domain. Therefore, an excess in bandgap over the value $T/300$ must be available if good electrolytic behavior is required over a wide range of chemical potentials. Curve II in Fig. 7.7 takes a safety factor of 100 in concentration into account.

7.1.5 Mixed Conductors with High Electronic Conductivity

In the usual reversible battery systems, electrode reactions are confined to the interface between the electrolyte and the electrode material. To achieve reaction in the depth of the material, porous structures with large surface area are

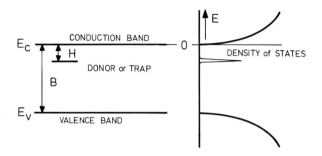

CONDUCTION BAND

E_C

H

DONOR or TRAP

B

E_V

VALENCE BAND

DENSITY of STATES

Fig. 7.8. Electronic energy band diagram and state density function (from *Heyne* [7.13])

required. Electronic conductors which also exhibit a high ionic conductivity seem to be promising materials for electrodes, because in such materials the electron transfer reaction need not occur only at the interface. Therefore we expect less electrode polarization, and, if the homogeneity range of the compound is broad, the reaction product will not separate as a second solid phase. This might be favorable for cyclic operation where mechanical stability and chemical reactivity must be maintained.

The requirements for a good electrode material have the consequence that both the electronic and the ionic charge carriers exhibit nonideal thermodynamical behavior. Thus most of the general statements deduced for electrolytes or compounds with small defect concentrations cannot be applied here. However, (7.18) remains valid with fair accuracy in many situations. Because it relates the experimentally accessible quantity μ_M^* with the chemical potential of electrons, the possibility of learning some details about the electronic band structure is open.

In the following considerations we assume that a metal excess is incorporated into the material according to the chemical reaction expressed in (7.15), or to an equivalent equation involving defects such as (7.3). This means that the metal excess results in an equivalent electron concentration. In the case of a metal deficit, this corresponds to an equivalent number of holes. Because the electronic system is degenerate, we must use Fermi instead of Boltzmann statistics to describe the distribution of the electrons over the available states. The occupation probability f of a certain electronic energy level (state) at energy eE_e is given by

$$f = 1/[1 + \exp e(E_e - F)/kT]. \tag{7.39}$$

The Fermi level F determines whether an energy state is substantially filled $(E_e < F)$ or empty $(E_e > F)$ and is completely equivalent to the electrochemical potential of electrons $\tilde{\mu}_e$ divided by e.

In Fig. 7.8 the well-known distribution of energy levels in a semiconductor is schematically represented, e.g. [Ref. 7.5, p. 27]. For the approximation of a spherical band structure [Ref. 7.6, Sect. 9.1], the states in the conduction and valence bands exhibit a density proportional to $E_e^{1/2}$. The zero-reference level for the electron energy is chosen at the band edge. If the same reference is used for the Fermi energy, the latter is equal to the thermodynamic potential of electrons μ_e, because the electrostatic potential term is then excluded.

The total concentration of electrons can be obtained by integration of the product of state density and probability of being occupied over electron energy

$$n = 2\pi^{-1/2} N_c \int_0^\infty [1 + \exp(\varepsilon - \eta)]^{-1} \varepsilon^{1/2} d\varepsilon$$
$$= 2\pi^{-1/2} N_c F_{1/2}(\eta). \tag{7.40}$$

N_c follows from (7.35), $\varepsilon = eE_c/kT$ and $\eta = eF/kT$; the function $F_{1/2}(\eta)$ can be obtained only by numerical integration and is available in tabulated form [7.14, 15]. For $\eta \ll 0$ it approaches $\frac{1}{2}\pi^{1/2} \exp\eta$, so that the Boltzmann formula is obtained.

$$n_{ideal} = N_c \exp\eta = N_c \exp(\mu_e/kT), \tag{7.41}$$

which corresponds to ideal thermodynamic behavior of the electrons. For nonideal thermodynamics the same expression holds for the "activity" a_n of the electrons

$$a_n = N_c \exp\eta = N_c \exp(\mu_e/kT), \tag{7.42}$$

so that using (7.40) the activity coefficient for this semiconductor model is known [7.16]. For holes generated by a metal deficit, equivalent reasoning leads to

$$p = 2\pi^{-1/2} N_v F_{1/2}(-\eta - eB/kT), \tag{7.43}$$

where the quantity B is introduced to keep the zero level at the lower conduction band edge as before (see Fig. 7.8).

In Fig. 7.9 the electron and hole concentration together with their activities are plotted against the Fermi level (expressed in units kT/e) on a logarithmic scale. At low concentrations the curves are straight lines in accordance with the Boltzmann approximation (7.41). These parts of the curves have been used in the Brouwer diagrams of Section 7.1.3. Deviations from the straight (activity) lines indicate the start of nonideality; this takes place at concentrations of about $2 \cdot 10^{19}$ cm^{-3}, where the Fermi level crosses the band edges. The distance between the two edge crossings on the horizontal scale corresponds to the bandgap. In the example of the figure it was chosen as 20 units kT/e or about 0.51 V at room temperature. Because the ion concentration is approximately $2 \cdot 10^{22}$ cm^{-3}, the deviation from stoichiometry that can generate such an electronic concentration is about 10^{-3} atomic fraction. This metal excess, expressed in atomic equivalents, equals the value $(n - p)$, which is also plotted. It passes through zero (minus infinity on the logarithmic scale) at the intrinsic point. Figure 7.9 holds for room temperature and for effective electron and hole masses equal to the free electron mass. Other temperatures or effective masses

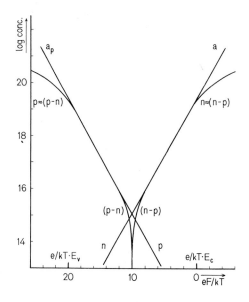

Fig. 7.9. Electron and hole concentrations and activities as a function of Fermi level. The curves for $(n-p)$ and $(p-n)$ correspond to metal excess or metal deficit concentrations (after *Heyne* [7.13])

only shift the vertical scale, because the density of states is then modified according to (7.35).

The graph is important because it relates practically accessible quantities. The Fermi energy (i.e., the electronic chemical potential) plotted along the abcissa can be determined as the emf of a suitable electrochemical cell, because it varies in parallel with μ_M^*, see (7.18). Furthermore, the deviation from stoichiometry δ can be varied by a known amount in such a cell by controlled electrolysis. Thus the effective density of electronic states is probed by such a "coulometric titration", and a value for the effective charge carrier mass can be calculated from the results [7.17–19].

The familiar shape of a titration curve becomes evident if the curve for $(n-p)$ is replotted on a linear scale for concentration, and at the same time the coordinate axes are interchanged, see Fig. 7.10.

However, it should be recognized that the theory given contains some inconsistencies at high deviations from stoichiometry. It was assumed that all excess metal atoms are completely ionized, so that there exists a quantitative correspondence between metal excess and free electron concentration (or between metal deficit and free hole concentration). But with a degenerate electronic system this is impossible, because then the Fermi level lies above the donor level (see Fig. 7.8) and consequently the latters occupancy becomes $<1/2$ [see (7.39)].

Moreover, at very high excess concentrations the one electron treatment on which the band picture and the effective mass approximation are based becomes questionable. The band structure is modified because an "impurity band" results from the overlapping donor level wave functions. Also the lower edge of the

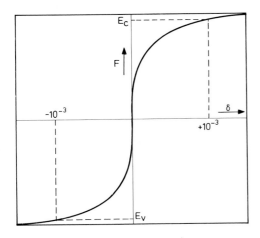

Fig. 7.10. Titration curve for a compound $M_{1+\delta}X$ on linear concentration scale. The band edges are passed at about 10^{19} at. cm^{-3} or 10^{-3} atom fraction (after *Heyne* [7.13])

conduction band becomes diffuse. Both effects lead to a concentration dependence of the effective mass [7.20, 21].

Although the theoretical description given loses validity, the coulometric titration method remains useful to determine the electrochemical potential and Fermi level as a function of deviation from stoichiometry, and thus to probe the electronic energy level distribution in disordered ionic compounds.

For the evaluation of experimental curves it is useful to note that at the inflexion point of a titration curve the electronic disorder is near intrinsic. If the bandgap is not very small, Boltzmann statistics may be used. From (7.41) and its counterpart for holes it follows that $(n-p)$ varies as a hyperbolic sine function with the deviation from stoichiometry. A simple calculation [7.22] shows that the intrinsic concentration n_0 and thus the bandgap can be found from the slope at the point of inflexion.

7.2 Charge Carrier Flow in Mixed Conductors

In this section a phenomenological description of the flow of charged species in solids will be given, without considering the mechanism of ionic mobility in general, nor the reasons for high mobility in solid electrolytes in particular. It will just be accepted that certain ion types are mobile, and the consequences for material and charge transport will be discussed.

Throughout this discussion isothermal conditions will be assumed, and the equations will be kept simple by considering only one-dimensional geometry and isotropic solids; coupling between fluxes will be neglected.

As is done in irreversible thermodynamics, we assume that in the systems considered, meaningful thermodynamic quantities can be defined, such as chemical and electrochemical potentials, although flows of material and charge are taking place. The gradient in the electrochemical potential is the driving force for the flow of charged particles.

7.2.1 Single Particle Equations

The current density carried by the particle species j can be written [Ref. 7.7, p. 59]

$$i_j = -\frac{\sigma_j}{z_j e}\,\text{grad}\,\tilde{\mu}_j. \tag{7.44}$$

The partial conductivity σ_j can be expressed in terms of the mobility u_j and the concentration c_j by

$$\sigma_j = c_j |z_j| e u_j. \tag{7.45}$$

The electrochemical potential $\tilde{\mu}$ can be split up into a chemical and an electrical part

$$\tilde{\mu}_j = \mu_j + z_j e\phi \tag{7.46}$$

where ϕ represents the electrostatic potential. Furthermore, activities can be defined from the following equation

$$\mu_j = \mu_j^0 + kT\,\ln a_j. \tag{7.47}$$

Using the equations above together with Einstein's relation which connects the mobility u_j with the particle diffusion coefficient D_j

$$D_j = (kT/|z_j|e)u_j, \tag{7.48}$$

it can be verified easily that Ohm's law for current transport without concentration gradient, and Fick's first diffusion law for material transport without electrical field are contained in (7.44) as limiting laws (the latter only for ideal behavior $c_j = a_j$). For the general case (7.44) can be transformed into the following forms

current density
$$i_j = -D_j c_j e z_j [\text{grad}\,\ln a_j + (z_j e/kT)\,\text{grad}\,\phi] \tag{7.49}$$

particle flux density
$$j_j = -D_j c_j [\text{grad}\,\ln a_j + (z_j e/kT)\,\text{grad}\,\phi]. \tag{7.50}$$

The latter equation can also be molded into the form of Fick's first law, putting $\text{grad}\,\phi = 0$

$$\begin{aligned} j_j &= -D_j c_j (\partial \ln a_j/\partial \ln c_j)\,\text{grad}\,\ln c_j \\ &= -D_j(\partial \ln a_j/\partial \ln c_j)\,\text{grad}\,c_j = -D_f\,\text{grad}\,c_j, \end{aligned} \tag{7.51}$$

where

$$D_f = (\partial \ln a_j/\partial \ln c_j)D_j. \tag{7.52}$$

Here the Fick's law diffusion coefficient D_f equals the particle diffusion coefficient D_j multiplied by a so-called "thermodynamic factor". The latter becomes unity for ideal behavior.

However, at this stage consideration of pure diffusion ($\operatorname{grad}\phi = 0$) is not fruitful and might even become misleading, because concentration gradients of one charged particle type cannot be applied without introducing either electric fields or concentration gradients for other species also. Moreover, the activity a_j is not only a function of c_j, but it depends also on the other concentrations c_i ($i \neq j$). Consequently, terms exhibiting other partial differential coefficients may occur, indicating cross-coupling of fluxes.

7.2.2 Simultaneous Diffusion of Several Types of Charge Carriers, emf of Galvanic Cells

We shall restrict our considerations to systems without a net *external* current, but where *internally* electronic and ionic charge carriers flow. The condition for cancellation of the latter currents is

$$\sum_k i_k = 0. \tag{7.53}$$

An unloaded galvanic cell is an example of such a system, and our analysis will lead to general expressions for the emf of such cells. But electrical connections are not essential, and the treatment applies as well to a sample of material in which gradients in electrochemical potential of one or more of its components are set up ("chemical diffusion"), for example, during equilibration after a change in the environment. The formation of solid reaction products, e.g., tarnishing layers by oxidation, is another practical situation to which the analysis applies.

When we use the transport number of species k defined by

$$t_k = \sigma_k / \sum_j \sigma_j, \tag{7.54}$$

(7.44) and (7.53) can be combined to yield

$$\sum_k (t_k/z_k) \operatorname{grad} \tilde{\mu}_k = 0. \tag{7.55}$$

To obtain expressions for potential differences measurable between two positions in an electrochemical system the quantity $\tilde{\mu}_k$ in (7.55) is split up into two parts. Traditionally this is done according to (7.46) (see for instance [7.1]). But in solid electrolytes μ_k of the mobile species cannot vary (7.14). Therefore, splitting $\tilde{\mu}_k$ according to the particle ionization equilibrium (7.16) is more prolific. For comparison, both calculation methods will be put side by side. Recall that a measured emf equals an integral of $d\phi$ only if the integration limits correspond to points in the same kind of material. The measured emf can also be considered as the (negative) integral of $(1/e)d\tilde{\mu}_e$ between the points. Then this restriction is not required.

Traditional for solutions | **Preferred for solids**

Splitting up of $\tilde{\mu}_k$ according to

$$\tilde{\mu}_k = \mu_k + z_k e\phi \qquad (7.56a) \mid \tilde{\mu}_k = \mu_k^* - z_k\tilde{\mu}_e \qquad (7.56b)$$

leads to

$$\text{grad}\,\phi = \frac{-1}{e}\sum_k \frac{t_k}{z_k}\,\text{grad}\,\mu_k \quad (7.57a) \left| \; \text{grad}\,\tilde{\mu}_e = \sum_k \frac{t_k}{z_k}\,\text{grad}\,\mu_k^* . \right. \qquad (7.57b)$$

Integration between positions indicated by ' and ", respectively, leads to expressions for the measurable voltage difference (emf)

$$E = \int_{'}^{''} d\phi = \frac{-1}{e}\int_{'}^{''}\sum_k \frac{t_k}{z_k}d\mu_k \quad (7.58a) \left| \; E = \frac{-1}{e}\int_{'}^{''} d\tilde{\mu}_e = \frac{-1}{e}\int_{'}^{''}\sum_k \frac{t_k}{z_k}d\mu_k^* . (7.58b) \right.$$

Thus we obtain two different expressions for the emf of a galvanic cell, which of course must lead to the same numerical value, although they differ in the use of ion chemical potentials in (7.58a), and in true chemical potentials of components, indicated by an asterisk in (7.58b).

Note, that in the summation a term for electrons is included in the a-equations, but not in the b-equations. This is so because there is no neutral equivalent of the charged electron, so that $\mu_e^* = 0$ [this follows from (7.56b) for index $k = e$].

To illustrate the use of the expression (7.58b), the emf of the following galvanic cell will be calculated.

$$\begin{array}{c|c|c} \text{Electrode I} & \text{Binary electrolyte} & \text{Electrode II} \\ \mu_M^{*'} & M^{z+}X^{z-} & \mu_M^{*''} \end{array} . \qquad (7.59)$$

We assume that in the electrolyte the charge carriers M^{z+}, X^{z-} and e^- (electrons) are present, while in both electrodes the chemical potentials of M are fixed, for example by definition of the partial pressures of X_2-gas. Equation (7.58b) then yields

$$E = -1/ze\int_{'}^{''}(t_M d\mu_M^* - t_X d\mu_X^*) = -1/ze\int_{'}^{''}(t_M + t_X)d\mu_M$$
$$= -1/ze\int_{'}^{''} t_{ionic}\,d\mu_M^* = -1/ze\int_{'}^{''}(1 - t_e)d\mu_M^* . \qquad (7.60)$$

The electronic transport number can thus be determined from the measurement of the emf of a galvanic cell.

Note that among the three transport numbers involved, t_M, t_X, and t_e, only one relation is found from the emf measurement; we also have $t_M + t_X + t_e = 1$, but still one relation is missing to allow separate determination of the two ionic transport numbers. This uncertainty arises because of the arbitrariness involved in the definitions of the separate fluxes and can be resolved only if the latter are defined with respect to a well-specified reference frame [7.23].

[handwritten: THAT'S NOT THE ONLY PROBLEM. THIS IS NOT A DILUTE SOLUTION]

7.2.3 Material Transport, Chemical Diffusion

We now continue our considerations of the combined movement of several types of charge carriers without external current. Such movement takes place when concentration gradients of the neutral species (or chemical potential gradients of the neutral components) are present in an ionic solid. Usually the transport of neutral defects is negligible in such compounds, as can be deduced from the correspondence between the ionic conductivity and the coefficient of self-diffusion D_s, in accordance with Einstein's relation (7.48),

$$\sigma_j = c_j |z_j| e u_j = c_j z_j^2 e^2 D_j / kT = N_j z_j^2 e^2 D_s / kT. \tag{7.61}$$

Here N_j stands for the total concentration of j-ions in the compound, the self-diffusion coefficient D_s describes the diffusive motion macroscopically, i.e., as if all ions (N_j) took part in the process, and not only the corresponding lattice defects with concentration c_j. Correlation effects are not considered here [7.24].

Usually, only one type of ion is mobile and thus we may restrict our considerations to the simultaneous flow of that ion type with electrons (or holes). The immobile ions can be used as the reference frame for the definition of motion and transport numbers.

There are two ways of (approximate) quantitative description of this simultaneous motion of charged particles leading to diffusion of a neutral species. They start from slightly different assumptions and also lead to somewhat different results. The first is due to C. Wagner [7.25–27] and was developed to describe the growth of "scaling" layers on metals. The second method is a description as "ambipolar" diffusion; it is extensively used in relation to semiconductors, electrolytic solutions and gas discharges [7.28].

We shall follow both methods and observe carefully where restrictive assumptions are introduced, and how they differ in the two cases. To do this, we deviate from the usual description found in the literature. To keep the formulas simple we shall make an extensive use of transport numbers (7.54).

Equation (7.44) and (7.55) are used as a start; they were derived under the condition of zero net electrical current, expressed in (7.53). With the use of (7.61) it is possible to write (7.44) in the form

$$j_j = -(c_j D_j / kT) \operatorname{grad} \tilde{\mu}_j, \tag{7.62}$$

where j_j is the flux density of particle type j. For two types of particles (7.55) takes the form

$$t_1 \cdot 1/z_1 \operatorname{grad} \tilde{\mu}_1 \qquad = -t_2 \cdot 1/z_2 \operatorname{grad} \tilde{\mu}_2.$$

This can also be written in the form

$$t_1(1/z_1 \operatorname{grad} \tilde{\mu}_1 - 1/z_2 \operatorname{grad} \tilde{\mu}_2) = -t_2 \cdot 1/z_2 \operatorname{grad} \tilde{\mu}_2 - t_1 \cdot 1/z_2 \operatorname{grad} \tilde{\mu}_2$$

$$= -1/z_2 \operatorname{grad} \tilde{\mu}_2. \tag{7.63}$$

Introducing the value for grad $\tilde{\mu}_2$ given in (7.63) into (7.62) for $j=2$ yields for the flux density

$$j_2 = -(z_2 c_2 D_2 t_1/kT)\,\mathrm{grad}\,(\tilde{\mu}_2/z_2 - \tilde{\mu}_1/z_1).\tag{7.64}$$

So far no restrictive assumptions have been introduced, and consequently the equations are valid for all types of compounds with two types of charge carriers.

The Wagner Treatment

We now take the way indicated by *Wagner* by assuming that *locally a chemical equilibrium between species is established*, which is not disturbed by the flow of particles. Let us define the mobile particles more closely as z-valent cations $(z=z_1>0)$ and electrons $(z_2=-1)$. The equilibrium condition then reads in correspondence with (7.17)

$$\mathrm{grad}\,\tilde{\mu}_j + z\,\mathrm{grad}\,\tilde{\mu}_e = \mathrm{grad}\,\mu_j^*.\tag{7.65}$$

Combination with (7.64) then yields for the electron flux density, which is equivalent to the (neutral) metal flux j_j^* passing through the material,

$$j_2 = j_n = z j_j^* = -nD_n t_j/(zkT)\,\mathrm{grad}\,\mu_j^*.\tag{7.66}$$

This is equivalent to Wagner's scaling equation, which can also be expressed in terms of conductivities,

$$j_j^* = -1/(z^2 e^2)\frac{\sigma_n \sigma_j}{\sigma_n+\sigma_j}\,\mathrm{grad}\,\mu_j^*$$
$$= -1/(z^2 e^2)t_n t_j \sigma_{tot}\,\mathrm{grad}\,\mu_j^*.\tag{7.67}$$

We now introduce the metal excess over the stoichiometric composition as a concentration c_j^*. This time the asterisk indicates the neutrality of the excess component, but it must be kept in mind that the actual species present are (interstitial) ions and electrons. We consider this excess as the diffusing item, and therefore try to mold (7.66) into the form of Fick's first diffusion law with a diffusion coefficient D^* (the "chemical" diffusion coefficient) and a gradient in c_j^*. Because μ_j^* is a unique function of c_j^* in a binary compound we may write

$$\mathrm{grad}\,\mu_j^* = (\partial\mu_j^*/\partial c_j^*)\,\mathrm{grad}\,c_j^*,\tag{7.68}$$

so that the flux equation becomes

$$j_j^* = -(nD_n t_j/z^2 kT)(\partial\mu_j^*/\partial c_j^*)\,\mathrm{grad}\,c_j^*.\tag{7.69}$$

Consequently, the chemical diffusion coefficent can be written

$$D^* = (t_j nD_n/z^2 kT)(\partial\mu_j^*/\partial c_j^*).\tag{7.70}$$

We again assume (see Sect. 7.1.5) that the metal excess has generated an equivalent electron concentration (e.g., full ionization of the metal interstitials) in accordance with the incorporation reaction

$$M_j^* \rightarrow M_j^{z+} + ze^-, \qquad (7.71)$$

so that

$$n = zc_j^*. \qquad (7.72)$$

The chemical diffusion coefficient thus becomes

$$D^* = t_j D_n (1/zkT)(\partial \mu_j^*/\partial \ln c_j^*), \qquad (7.73)$$

$$D^* = t_j D_n (1/zkT)(\partial \mu_j^*/\partial \ln n). \qquad (7.74)$$

This quantity is the effective diffusion coefficient which is common to the ionic defects and the electrons, and which also describes the speed with which a deviation from stoichiometry moves through the material. *The driving force is a gradient in concentration of a neutral species in all cases.*

The last two factors in (7.73) or (7.74) form a "thermodynamic factor" t.f. It can be written in several equivalent forms:

$$\begin{aligned}
\text{t.f.} &= (1/zkT)(\partial \mu_j^*/\partial \ln c_j^*) \\
&= (1/zkT)(\partial \mu_j^*/\partial \ln n) \\
&= (1/z)(\partial \ln a_j^*/\partial \ln n) \\
&= (1/z)(\partial \ln a_j^*/\partial \ln c_j^*). \qquad (7.75)
\end{aligned}$$

For nondegenerate electrons $kT\partial \ln n = \partial \mu_e$ so that

$$\text{t.f.} = (1/z)(\partial \mu_j^*/\partial \mu_e). \qquad (7.76)$$

In compounds with high disorder (7.18) holds, so that $\partial \mu_j^*/\partial \mu_e = z$ and consequently the thermodynamic factor becomes unity. Thus we obtain for *electrolytes and mixed conductors with a nondegenerate electronic system,*

$$D^* = t_j D_n. \qquad (7.77)$$

If (7.18) does not hold very accurately, as is the case in situations represented by mixed disorder regions in the Brouwer diagrams of Figs. 7.1 and 7.2 or by regions near them, it follows from (7.17) that $\partial \mu_j^*/\partial \mu_e$ is larger than z, so that the chemical diffusion coefficient is also somewhat in excess of $t_j D_n$.

Equation (7.77) shows that the electronic diffusion coefficient (mobility) is the determining quantity for the speed of chemical diffusion in electrolytes

$(t_j=1)$. Because the electronic mobility is high in comparison with defect mobilities, the chemical diffusion in ionic conductors is very fast compared to tracer diffusion. If this causes surprise, it is well to realize that, when a chemical diffusion experiment is carried out, not only is the defect concentration raised at a certain position, but also the electronic concentration is increased correspondingly. The electronic charge carriers are the most mobile ones and try to move fast. In the case of dominating ionic conductivity, a retarding field that might keep the electrons from fast movement to places of lower concentration cannot be built up. Thus the fast electrons determine the speed of the common diffusion process.

From the definition of transport numbers (7.54) and (7.61) it can be seen that

$$nD_n t_j = z^2 c_j D_j t_n .\tag{7.78}$$

Thus the expression for the chemical diffusion coefficient (7.73) can also be written

$$D^* = t_n D_j (c_j/c_j^*)(1/kT)(\partial \mu_j^*/\partial \ln c_j^*),\tag{7.79}$$

when (7.72) is used also. Introducing the coefficient of self-diffusion

$$D_s = (c_j/N_j)D_j ,\tag{7.80}$$

we obtain

$$D^* = t_n D_s (N_j/c_j^*)(1/kT)(\partial \mu_j^*/\partial \ln c_j^*).\tag{7.81}$$

Here

$$c_j^*/N_j = \delta\tag{7.82}$$

equals the relative deviation from stoichiometry as usually employed in the chemical formula.

Note that for cation-disordered sublattice compounds, $N_j=c_j$ so that the defect diffusion coefficient D_j equals the self-diffusion coefficient D_s.

The final expression (7.81) for the chemical diffusion coefficient can be experimentally verified. For example, in compounds with dominating electronic conductivity $(t_n=1)$ both $\partial \mu_j^*/\partial \ln c_j^*$ and δ can be found from a coulometric titration (see Sect. 7.1.5). D^* and D_s can be found from a chemical and a tracer diffusion experiment, respectively. Alternatively, the latter can be replaced by a measurement of the ionic partial conductivity, see (7.61) [7.29].

In (7.81) the large value of D^* with respect to D_s stems mainly from the factor $N_j/c_j^* = 1/\delta$, which is usually large. The thermodynamic factor we use here [the product of the last two factors in (7.81)] equals z times that defined before in (7.75), and consequently is near unity for ionic conductors, for which we thus have

$$D^* = t_n D_s (z/\delta).\tag{7.83}$$

Note that for this case (7.77) remains also valid. Therefore, although D^* is much larger than D_s, it remains smaller than D_n.

Note that it is common practice to include the factor z/δ in the thermodynamic factor in other definitions of the latter [7.30–32]. It is even possible to include the electrical interaction in it [7.33].

We recapitulate the restrictions introduced in the derivation of the expression for the chemical diffusion coefficient, (7.77): 1) Local equilibrium between moving species is not disturbed. 2) Ionic disorder resulting from a deviation from stoichiometry is negligible with respect to the natural disorder present [validity of (7.18)]. 3) The electron concentration is coupled to the metal excess by the simple incorporation mechanism of (7.77) and (7.72). 4) The concentration of electronic charge carriers is so low that they behave ideally, i.e., Boltzmann statistics can be applied to them.

7.2.4 Relation Between Chemical and Ambipolar Diffusion

In ambipolar diffusion theory [7.28] it is assumed that both carriers obey Fick's first diffusion law in the absence of an electric field. This means that ideal thermodynamic behavior is postulated [see (7.52)]. When we correspondingly replace $\tilde{\mu}_2$ and $\tilde{\mu}_1$ in (7.64) by logarithmic terms in concentration, we obtain, because the electrostatic potential contributions of $\tilde{\mu}_1$ and $\tilde{\mu}_2$ cancel,

$$j_2 = -z_2 c_2 D_2 t_1 [(1/z_2 c_2)\operatorname{grad} c_2 - (1/z_1 c_1)\operatorname{grad} c_1]. \tag{7.84}$$

In addition, the assumption of "quasi-neutrality" is made. On one hand this means that the particles are forced to flow together by an attractive electrostatic field which is formed by a slight misadjustment of the charge balance of the moving particles, so that they have their diffusion coefficient in common. On the other hand, the concentration misadjustment is so small, relatively, that in all considerations other than those concerned with space charge, the concentrations may be taken as equal (in terms of equivalents).

This quasi-electroneutrality condition holds well for sample dimensions large compared with a Debye length. This means that the higher the charge carrier concentration, the more accurate is the approximation.

Quantitatively it can be expressed in our case as

$$z_1 \operatorname{grad} c_1 + z_2 \operatorname{grad} c_2 = 0. \tag{7.85}$$

Combination with (7.84) yields

$$j_2 = -z_2^2 c_2 D_2 t_1 [(1/z_2^2 c_2) + (1/z_1^2 c_1)]\operatorname{grad} c_2. \tag{7.86}$$

The common diffusion coefficient D_a thus becomes

$$D_a = t_1 D_2 (z_1^2 c_1 + z_2^2 c_2)/(z_1^2 c_1). \tag{7.87}$$

This can be modified into either

$$D_a = \frac{(c_1 z_1^2 + c_2 z_2^2) D_1 D_2}{c_1 z_1^2 D_1 + c_2 z_2^2 D_2} \tag{7.88}$$

or

$$D_a = t_j D_n + t_n D_j. \tag{7.89}$$

In the last equation the two particle types are identified again with electrons and defects of type j.

The similarity of the last equation with that resulting from the previous treatment, (7.77), is striking, especially if it is recognized that in the many cases of interest to us the second term is very small.

We finally compare the assumptions involved here with those made in the Wagner type of treatment. In addition to ideality of electronic carriers, here also ideal behavior of the ionic defects was postulated. On the other hand, because local equilibrium is irrelevant in the last type of treatment, it has found a wide application in connection with semiconductors, photoconductors and gas discharges, where large deviations from equilibrium are very common. It seems that instead the condition of quasi-neutrality couples the two types of carrier concentrations.

However, a closer examination of the two lines followed in the derivation of (7.77) and (7.89) shows that the two methods are more closely connected than is usually assumed. Especially the condition of local equilibrium, which was used in an early stage in Wagner's treatment [in our version of this calculation in the transition from (7.64) to (7.66)], appears to be premature.

In the following, we shall reconsider the derivations for the two expressions for the chemical diffusion coefficient. But in doing this, we alter the order in which the various assumptions are introduced in such a way that the restrictions common to both treatments come first. This concerns the ideality of electronic carriers and the incorporation reaction (7.71).

We start again with the flux equation (7.64), which has general validity, and in which the two electrochemical potentials may be replaced by chemical ones as before.

$$j_n = -(n D_n t_j / kT) \operatorname{grad}(\mu_e + \mu_j / z). \tag{7.90}$$

Making use of electron ideality, (7.41), and the incorporation balance, (7.72), this can be transformed into

$$z j_j^* = j_n = -t_j D_n (z + \partial \mu_j / \partial \mu_e) \operatorname{grad} c_j^*. \tag{7.91}$$

Here it is assumed again that μ_j is a function of c_j^* only. The effective diffusion coefficient thus becomes

$$D^* = t_j D_n [1 + 1/z (\partial \mu_j / \partial \mu_e)]. \tag{7.92}$$

This formula must still contain all expressions for D^* and D_a derived before.

The factor between brackets represents a thermodynamic factor. However to transform it to the form of (7.76), we must make use of the ionization equilibrium expressed in (7.16) and (7.17). This equilibrium condition is *not* required for ionic conductors, because in that case μ_j is constant [see (7.14)], and consequently the factor in brackets becomes unity. In that case (7.77) is found again.

Another relation between μ_j and μ_e, with which the term in brackets can be evaluated, follows from the quasi-neutrality condition $dn = z\,dc_j$ used in the ambipolar approximation method. But to transform the concentrations from the neutrality condition into chemical potentials we need, the additional assumption of ideality (or a specified form of nonideality) is required. In that case $\partial\mu_j/\partial\mu_e$ can be evaluated as

$$\partial\mu_j/\partial\mu_e = \partial\ln a_j/\partial\ln a_e = \partial\ln c_j/\partial\ln n$$
$$= (n/c_j)(\partial c_j/\partial n) = n/(zc_j), \qquad (7.93)$$

so that using also (7.54) and (7.61) in addition to (7.92) the ambipolar formula (7.89) is reproduced.

These considerations show that the two methods of approach are essentially similar, and that only in the final steps do deviations occur. These differences act only on the second term of the expressions for the effective diffusion coefficient, which forms the smaller part for compounds of interest here. The assumption of local equilibrium is required only if (7.74) or one of its equivalents is used for the chemical diffusion coefficient. Equation (7.92) is more general however.

7.2.5 Influence of Electronic Trapping

In the previous subsection it was shown that the chemical diffusion coefficient of electrolytes or mixed conductors is of the order of magnitude of the large electronic diffusion coefficient. So we expect values near $1\,\mathrm{cm^2/s}$. The time constant of equilibration processes with a surrounding atmosphere is controlled by this diffusion coefficient, and thus we expect equilibration times of the order of seconds for samples of centimeter size. In practice much lower values are often found, especially for electrolytes.

A possible explanation is that not all of the electronic carriers formed by the incorporation of a metal excess (or deficit) are present in the state of "free" electrons (or holes), but that they are largely in states bound to defects or impurity centers. This mechanism is called "trapping". It can also be described as the reduction (or oxidation) of impurity atoms to a different valence state or as the formation of color centers [7.34, 35]. The result of an electron trapping process can be that an originally neutral center becomes negative, but also that an originally positive center becomes neutral. Examples of possible trapping processes relevant for ZrO_2 or alkali (silver) halides are, given in Kröger and Vink notation,

$$Ti_{Zr}^* + e^- \to Ti_{Zr}^- \qquad\qquad V_{Cl}^+ + e^- \to V_{Cl}^*\ (\text{F-center})$$
$$Fe_{Zr}^{2-} + h^+ \to Fe_{Zr}^- \qquad\qquad Ag_i^+ + e^- \to Ag_i^* .$$

It must be stressed that here, and in what follows, trapping will be considered as a reversible reaction between point defects and electrons, while the electronic system, consisting of free and trapped electrons and holes, is in thermodynamic equilibrium locally. This is in contrast with the trapping phenomena usually considered in photoconductors [7.36, 37] or semiconductors, where the electron-hole equilibrium is disturbed by *injection* [7.38] from electrodes or junctions. There the distribution of electronic carriers over energy levels between the bands (traps, recombination levels, etc.) may strongly deviate from the equilibrium distribution. This different use of the trapping concept must be kept in mind.

To incorporate trapping in our considerations about chemical diffusion, we must modify the electroneutrality condition (7.7) by the addition of a species which can have two states of charge. The concentrations of empty and filled centers then follow from a Kröger and Vink analysis and can be incorporated in a Brouwer diagram [7.35]. An alternative, simple way of description, in which the need for specification of the exact nature of the center is avoided, is to divide the electrons or holes into free and bound (trapped) ones. To be definite we consider electrons (concentration n) which can become trapped in neutral centers (concentration N_t) to give filled traps of negative charge (concentration n_t).

Now we assume that deviations from stoichiometry are so small that we stay in a situation of ionic disorder, where neither the free nor the trapped electrons dominate the neutrality equation. So we stay in the same region of the Brouwer diagram characterized by the limiting neutrality equation (7.12) or (7.13). The only change is in the incorporation reaction (7.71), which must lead to the following relation between metal excess and electronic concentrations

$$zc_j^* = n + n_t \tag{7.94}$$

instead of (7.72).

A relation between n and n_t follows from the Fermi formula (7.39) which, when applied to both the conduction band (with effective state density N_c) and the trapping level (with state density N_t), leads to

$$n_t = N_t n \left[N_c \exp(-eH/kT) + n \right]^{-1}. \tag{7.95}$$

Here H is the depth of the trapping level below the conduction band edge (see Fig. 7.8).

The Boltzmann approximation is used for the conduction band. For low excess metal concentration, which is usual in electrolytes, n can be neglected with respect to $N_c \exp(-eH/kT)$ (the Fermi level is then below the trap level and most traps are empty) so that we obtain

$$n_t/n = (N/N_c) \exp(eH/kT) = \alpha - 1. \tag{7.96}$$

The ratio between the concentrations of trapped and free carriers is thus constant; this constant is put equal to $(\alpha - 1)$ and is called the trapping factor. Consequently, the incorporation balance (7.94) becomes

$$zc_j^* = \alpha n. \tag{7.97}$$

In the large bandgap materials we consider here, usually $H > e/kT$, so that α can be a very large number.

If we follow the derivations of the formulas for the chemical diffusion coefficient given in the previous subsection again, but replace (7.72) each time it is used by (7.97), we obtain

$$D^* = t_j D_n / \alpha \tag{7.98}$$

and

$$D_a = t_j D_n / \alpha + t_j D_p. \tag{7.99}$$

We see that the effective diffusion coefficient is lowered by the large factor α as a result of trapping. This could be expected because, most electrons being bound to the trapping centers, the average electron diffusion coefficient is strongly reduced.

Another way to look at the situation is to state that, at a certain chemical potential of electrons (Fermi level), resulting from a certain component partial pressure, the concentration of electrons in the conduction band is the same, but the corresponding deviation from stoichiometry is increased by the trapping factor α, because most of the excess electrons are bound by the traps. This larger excess must be moved by the same number of free electrons as in the trap-free situation. Consequently, the chemical diffusion process is slowed down.

The low chemical diffusion coefficient of oxygen in stabilized zirconia, which corresponds to a low hole mobility, was explained by such a trapping mechanism for holes [7.10, 39]. The strong temperature dependence of the experimental hole diffusion coefficient is consistent with this explanation, because α shows the temperature dependence expressed by (7.96). The activation energy of the hole mobility is a measure for the depth of the trapping level. Because a study of the chemical diffusion coefficient of oxygen as a function of impurity content is not available, the nature of the trapping centers in samples of this compound is unknown.

In compounds with a very high electronic conductivity, the condition for validity of (7.96), viz., that most of the traps remain empty, is not fulfilled. To the contrary, if the Fermi level is substantially above the level of the traps, the latter are practically all filled, and changes in excess metal concentration lead to equivalent changes in free electron concentration in accordance with the differential form of the original incorporation balance (7.72): $dn = z \, dc_j^*$. The number of trapped electrons cannot be changed any more. For such compounds equilibration with a surrounding atmosphere or with a solid in contact with it is very fast. The chemical diffusion coefficient has the high value to be expected for free electronic carriers [corrected with t_j, see (7.77)], notwithstanding the possible presence of impurities with potential trapping capabilities. Examples are the values found for Ag_2S and Ag_2Se [7.29, 40].

7.3 Experimental Methods to Separate Ionic and Electronic Conductivity Parameters

In this section, some experimental methods by which conductivities and mobilities of electronic charge carriers can be determined in compounds exhibiting a mixed conductivity mechanism will be discussed. Normal methods to determine the total conductivity of samples will not be considered. Although the possibilities of several types of electrochemical cells will be studied, an exhaustive review of techniques in which solid electrolytic cells can be used for material characterization will not be given (see Sect. 6.4). Instead, we shall concentrate on some examples in which small partial conductivities (especially of electronic nature) can be measured, and the special effects such a minor conductivity mechanism might cause. In these considerations the theory developed in the first two section will be applied. As has been shown by C. Wagner [7.41], galvanic cells with special combinations of reversible or partially blocking electrodes can be used for the evaluation of mixed conductor properties.

7.3.1 emf Method of Transport Number Determination

In this method the sample to be studied is placed between two reference electrodes which are supposed to keep the chemical potential at fixed values at the two interfaces see (7.59, 60). However, if the transport number of electrons is not zero, (7.67) shows that a material flow takes place. This causes extraction of the mobile component from one electrode and supply to the other. Resulting deviations from the equilibrium chemical potentials then may lead to over-estimation of the electronic transport number. In practice it is often difficult to check whether the reversibility of the electrodes is sufficient [7.42]. A possible way to test, and correct, is to use samples with different thicknesses to modify the material flux.

A more fundamental difficulty is that the electronic transport number is not a constant, varying with composition. This is the reason that in the expression for the emf of a solid state cell (7.60), the transport number t_{ion} cannot be taken in front of the integral sign. Nevertheless this is sometimes done and in this way a kind of average transport number defined. This has a limited meaning, however, and has been the origin of some confusion in the literature [7.43].

To obtain meaningful transport numbers, only a small chemical potential (partial pressure) difference may be applied, so that the emf formula can be used in the differential form. To study the variation of t_{ion} with partial pressure the situation at both electrodes must be varied simultaneously, which may present experimental problems.

An alternative method is to keep one electrode system constant and vary only the chemical potential of the other electrode. Differentiation of the integral of (7.60) with respect to the upper integration limit then yields

$$\left(\frac{\partial E}{\partial \mu_M^{*''}}\right)_{\mu_M^{*'} = \mu_M^*} = (-1/ze)\, t_{ion}, \tag{7.100}$$

or

$$\left(\frac{\partial E}{\partial \ln P_M''}\right)_{P_M' = P_M} = (-kT/ze)\, t_{ion}. \tag{7.101}$$

Thus the ionic transport number at chemical potential μ_M^* (or partial pressure P_M) can be found from the slope of a plot of E vs $\mu_M^{*''}$ at a constant $\mu_M^{*'}$.

Another possibility to obtain full transport number information from a limited number of measurements is to make use of the functional dependences derived for the electronic conductivities given in Section 7.1. Equations (7.28a) and (7.29a) give the dependence of σ_n and σ_p on partial pressure of component X, so that, using the definition of transport numbers given in (7.54), we may write

$$\sigma_e = \sigma_n + \sigma_p = \sigma_e^0 \left[(P_0'/P_X)^{1/y} + (P_X/P_0')^{1/y} \right] \tag{7.102}$$

and

$$
\begin{aligned}
t_{ion} &= \sigma_i/(\sigma_i + \sigma_e) = (1 + \sigma_e/\sigma_i)^{-1} \\
&= \{1 + (\sigma_e^0/\sigma_i)\left[(P_0'/P_X)^{1/y} + (P_X/P_0')^{1/y}\right]\}^{-1}.
\end{aligned}
\tag{7.103}
$$

As before σ_e^0 is the electron or hole conductivity at the intrinsic point (where $n = p$) and P_0' represents the partial pressure of X where this occurs.

By (7.103) t_{ion} is expressed as a function of P_X using as parameters the material constants σ_e^0/σ_i and P_0'. Alternative material constants are the X-partial pressures at the electrolytic domain boundaries: P_n and P_p.

The relations between these parameters and those used in (7.103) are given by the relations (7.31a) obtained earlier. Using them, (7.103) can be transformed into

$$t_{ion} = [1 + (P_n/P_X)^{1/y} + (P_X/P_p)^{1/y}]^{-1}. \tag{7.104}$$

Integration of (7.60) using this value for t_{ion} and replacing $d\mu_M^*$ by $-kT\, d\ln P_X$ leads, with a slight approximation allowed if $P_p \gg P_n$, to

$$E = \frac{kT}{e}\left(\ln \frac{P_p^{1/y} + P_X'^{1/y}}{P_p^{1/y} + P_X''^{1/y}} + \ln \frac{P_n^{1/y} + P_X''^{1/y}}{P_n^{1/y} + P_X'^{1/y}} \right). \tag{7.105}$$

In this relation, which was first derived by *Schmalzried* [7.44] for oxidic electrolytes, P_X'' and P_X' are the partial pressures at the two electrodes.

Matching of this expression with practical results allows determination of P_p and P_n, and consequently, by the use of (7.104), the ionic transport number at any partial pressure P_X. However, for values of P_X'' and P_X' in the range $P_n \ll P_X \ll P_p$, the emf given by (7.105) approaches the Nernst equation

$$E = (kT/ye)\ln(P_X''/P_X'), \tag{7.106}$$

so that no information about transport numbers can be obtained. Practical results can be obtained only if one of the electrode chemical potentials is near either P_n or P_p. Knowledge of the domain boundary pressures, together with the ionic conductivity, allows calculation of σ_e and P_0' from (7.31a). Knowing σ_e the electronic bandgap follows from (7.34–36), when suitable values for mobility and effective mass are introduced. This method of determination of the bandgap from electrochemical data was applied to stabilized zirconia and gave results in agreement with the optical properties [7.13, 39].

The method has been used extensively to determine the lower electrolytic domain boundary for stabilized zirconia and related compounds. Because in zirconia P_p is very high [7.39, 45, 46] the first term in brackets is near zero at practical values of P_{O_2} at the electrodes. Choosing P_{O_2}'' in the normal pressure range (e.g., that of air) and P_{O_2}' very low ($H_2 + H_2O$ mixture [7.47, 48] or base metal + metal oxide [7.49]), the second logarithmic term becomes sensitive to P_n and this quantity can be obtained from comparison with measured values.

At P_{O_2}' values much lower than P_n, E becomes independent of P_{O_2}'

$$E_{max} = (kT/4e)\ln(P_{O_2}''/P_n). \tag{7.107}$$

This maximum potential can be generated by chemical means, but also by having the emf measurement preceded by an electrolysis in a situation with no oxygen present at the cathode. The electrolyte itself is then reduced at the cathode to a state with dominant electron conductivity. After interruption of the current, the emf stabilizes on E_{max} [7.48, 50].

Results of the above-mentioned measurements are reviewed by *Fischer* and *Jahnke* [7.51], by *Goto* and *Pluschkell* [7.52] and by *Etsell* and *Flengas* [7.53].

A source of error in measurements of this type may be formed by the penetration of the active component through the electrolyte, which is coupled by its electronic conductivity. Although the influence of the latter on the emf is well taken care of by the formulas given, the extremely low partial pressures required at one of the electrodes can easily be influenced by a small material flow.

7.3.2 Determination of Small Electronic Transport Numbers

The material transport resulting from a chemical potential gradient, which was shown to perturb a transport number determination by emf measurement, can itself be used to obtain information about an electronic conduction process so

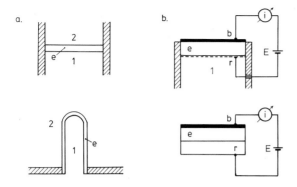

Fig. 7.11a and b. Schematic setups for, (a) permeation and (b) polarized cell techniques. e is the electrolyte, separate gas volumes are indicated by 1 and 2, b is a blocking and r a reference electrode. i is a current meter, E the polarization voltage source

small that the corresponding deviation from Nernst's law cannot be observed. Especially for sintered oxides this method is simple and sensitive, because it is essentially a measurement of oxygen leak-through.

In another method to determine small electronic conductivity components, galvanic cells containing one reference and one inert blocking electrode [7.41] are used. A voltage is applied in such a direction that the ion current is blocked; in the polarized state the residual current is carried by electronic carriers exclusively.

The experimental setups of the two methods are entirely different as shown in Fig. 7.11. In the permeation method two gas spaces are present with known partial pressures, and means are provided to analyze the gas evolution at the low pressure side, either through a flow system [7.54], or by vacuum pressure gauges [7.39]. Electrodes and electrical circuitry are not required, but may be present. On the other hand, the polarization method relies completely on electrical voltage and current measurements.

Notwithstanding this difference, there is a great similarity in the fundamental aspects of the two methods. In both cases a chemical potential gradient is present in the sample. In the permeation cell this gradient is obtained by the application of different gas atmospheres, and, for small t_e, a Nernst emf builds up which can be measured if electrodes are present. In the polarized cell the partial pressure is fixed only at the reference electrode. At the blocking electrode a certain chemical potential is generated as a result of the application of an electrical potential difference. The chemical potential (partial pressure) is built up by a transient electrolysis current, and is in the final state coupled with an emf which exactly counterbalances the applied voltage.

At small t_e the ionic flow in the permeation cell is very low; in the polarized cell it is exactly zero if the blocking action of the inert electrode is perfect. Thus in both cases grad $\tilde{\mu}_j \approx 0$. Because in ionic conductors μ_j is fixed (7.14) the electrical field strength grad ϕ must also be negligible [from (7.46)], and consequently, the electronic flux, which is equivalent to the permeation rate, must be a pure diffusional flow. Note that in both cases, suppression of the ionic current is present only in a stationary state where the concentration and composition gradients have attained stable values. Both methods will now be considered separately in a more detailed way.

7.3.3 The Permeation Technique (Static)

Because of the importance of this method for oxides we continue its discussion in terms of oxygen. The steady state permeation rate can be found from the flux equations given in the discussion of chemical diffusion in Section 7.2.3. We introduce $kT \ln P$ for μ_j^* (for typographical reasons P will be used for P_{O_2} in this subsection), put $z=4$ and express the oxygen flux in terms of the equivalent electrical current i. Equation (7.67) then yields

$$i=(kT/4e)t_e\sigma_{ion} \,\mathrm{grad}\ln P. \tag{7.108}$$

Note that in this stage we do not make any assumption about the magnitude of the electronic transport number. Integration over the sample thickness L leads to

$$i=(kT/4eL)\sigma_{ion}\int_{'}^{''} t_e d\ln P. \tag{7.109}$$

Assume that we have found from experiments a set of values of the permeation rate i at different oxygen pressures P'' at the high pressure side of the permeation cell, but at a fixed value P' at the low pressure side. For the evaluation of the transport number t_e at a certain oxygen pressure P from this set of data, it is useful to differentiate (7.109) with respect to the upper integration limit,

$$(\partial i/\partial \ln P'')_{P''=P}=(kT/4eL)\sigma_{ion}t_e. \tag{7.110}$$

By means of this relation t_e, the transport number at P can be expressed in terms of the slope of the experimental i vs $\ln P''$ plot at the pressure value P.

This analysis contains no restrictions and can be used when nothing whatsoever is known about the σ_e dependence on oxygen pressure, e.g., when it is not known whether the electronic conductivity is n- or p-type or near intrinsic in the pressure range involved.

Often some knowledge is available beforehand. For instance, t_e may be known to be small, so that we may put $t_e\sigma_{ion}=t_e\sigma_{tot}=\sigma_e$ in the preceding expressions. Moreover, we may introduce the oxygen pressure dependences of σ_n and σ_p given by (7.28a) and (7.29a). We do this in the form

$$\sigma_e=\sigma_n+\sigma_p=\sigma_{n,1}P^{-1/4}+\sigma_{p,1}P^{1/4}.$$

Here $\sigma_{n,1}$ and $\sigma_{p,1}$ are the electron and hole conductivities at 1 atm. Integration of (7.109) with these substitutions yields, also using (7.45) and (7.48),

$$i = (kT/eL)[-\sigma_{n,1}(P''^{-1/4} - P'^{-1/4}) + \sigma_{p,1}(P''^{1/4} - P'^{1/4})] \qquad (7.111)$$

$$i = (kT/eL)(-\sigma_n'' + \sigma_n' + \sigma_p'' - \sigma_p')$$
$$= (e/L)[(n'-n'')D_n + (p''-p')D_p]. \qquad (7.112)$$

Here the primed conductivities represent the electronic partial conductivities of electrons and holes at the two oxygen pressures maintained at the boundaries.

Simplification results when one of the boundary pressures is very low, so that $P''^{1/4} \gg P'^{1/4}$ and consequently $P''^{-1/4} \ll P'^{-1/4}$. We obtain

$$i \approx (kT/eL)(\sigma_{n,1}P'^{-1/4} + \sigma_{p,1}P''^{+1/4}) \qquad (7.113)$$

$$i \approx (kT/eL)(\sigma_n' + \sigma_p'') = e/L \quad (n'D_n + p''D_p). \qquad (7.114)$$

The permeation rate is dominated by the larger of the two boundary conductivities. If the lower pressure P' is very low, so that $\sigma_n' > \sigma_p''$, the permeation is controlled by the n-type conductivity at the low pressure side. In that case the permeation rate is proportional to $P'^{-1/4}$. If such an extremely low partial pressure is not maintained at the low pressure side, but a value near the intrinsic pressure or higher is used, we have the opposite case. Then the p-type term prevails and the permeation rate varies in proportion to $P''^{+1/4}$, and is independent of the pressure at the low pressure side. Often this situation has been tacitly assumed to be present. The analysis showed that, if the pressure dependence fits the $P''^{+1/4}$ law, this is justified. Lower pressures can however also be expected. Sometimes a higher power dependence has been found. This can be taken as an indication that an interface reaction, rather than the chemical diffusion, is partially rate determining. It can be expected that such rate limitations come into play especially when the electronic conductivity of the material under test is not low.

Results of these type of measurements have been recently reviewed by *Fischer* and *Jahnke* [7.51] and by *Fouletier* et al. [7.55].

7.3.4 The Polarized Cell Technique (Static)

In the steady state of a polarized cell no ionic current is flowing, so that $\mathrm{grad}\,\tilde{\mu}_i = 0$ if i refers to the mobile ionic species [see (7.44)]. From (7.65) it follows that in local equilibrium

$$\mathrm{grad}\,\mu_i^* = z_i\,\mathrm{grad}\,\tilde{\mu}_e. \qquad (7.115)$$

The electronic current, which also follows from (7.44), this time written with index $_e$ and $z_e = -1$, thus becomes

$$i_e = (\sigma_e/e)\,\mathrm{grad}\,\tilde{\mu}_e = [\sigma_e/(z_i e)]\,\mathrm{grad}\,\mu_i^*, \tag{7.116}$$

where σ_e is a function of μ_i^*. Integration over the sample thickness yields for the current flowing in the outer circuit in the final stationary state

$$\begin{aligned}
i_e &= 1/(z_i eL) \int'' \sigma_e d\mu_i^* \\
&= kT/(z_i eL) \int'' \sigma_e d\ln P_i.
\end{aligned} \tag{7.117}$$

Note the similarity with (7.109) for the permeation cell. If in the latter $\sigma_{\mathrm{ion}} = \sigma_{\mathrm{total}}$ the equations become identical (for $z_i = 4$). But in the case considered here only one of the electrode chemical potentials is fixed, e.g., $\mu_i^{*'}$ (or P_i'). The other $\mu_i^{*''}$ is determined by the applied potential difference E. Its value follows directly from integration of (7.115):

$$E = -1/e \int'' \tilde{\mu}_e = -1/(z_i e) \int'' d\mu_i^* = -(\mu_i^{*''} - \mu_i^{*'})/(z_i e). \tag{7.118}$$

With (7.18) we can also write

$$E = -(\mu_i'' - \mu_e')/e = -(kT/e)\ln(n''/n') = (kT/e)\ln(p''/p'). \tag{7.119}$$

Differentiation of (7.117) with respect to the upper integration limit gives

$$(\partial i_e/\partial \mu_i^{*''})_{\mu_i^{*''} = \mu_i^*} = 1/(z_i eL)\sigma_e, \tag{7.120}$$

which can also be written according to (7.118) [7.41]

$$(\partial|i_e|/\partial E)_{\mathrm{at}\,E} = \sigma_e/L. \tag{7.121}$$

If a set of experimental data is available in the form of a plot of the steady state polarization current as a function of polarization voltage, the electronic conductivity at a chemical potential corresponding to E can be found from the slope of this curve at the value E.

More detailed information about the shape of such a plot can be found using the dependences of σ_n and σ_p on chemical potential. These follow from the coupling of μ_e with μ_i^* expressed by (7.18) and the relation between concentration and chemical potential for nondegenerate electrons and holes, which is of the usual form (7.5). Assuming constant mobilities of the electronic charge carriers, we may write

$$\sigma_e = \sigma_n + \sigma_p = \sigma_n' \exp(-e\mu_i^*/kT) + \sigma_p' \exp(e\mu_i^*/kT). \tag{7.122}$$

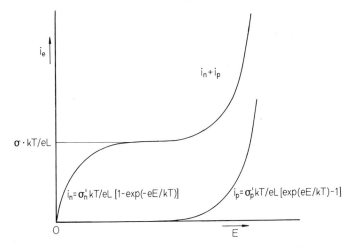

Fig. 7.12. General shape of steady state current of a polarized cell. The electrolyte passes from n-type, in equilibrium with the parent metal, to p-type electronic conductivity

Here σ'_n and σ'_p represent the electron and hole conductivities in a situation with a chemical potential determined by the reference electrode. This chemical potential is taken as the zero point in the expression above. Application of (7.118) and integration of (7.117) with (7.122) inserted leads to the following expression derived by C. *Wagner* [7.41]

$$|i_e| = kT/eL \, \{\sigma'_n[1 - \exp(-eE/kT)] + \sigma'_p[\exp(eE/kT) - 1]\} \,. \tag{7.123}$$

The general shape of a curve according to this expression is shown in Fig. 7.12. The electronic conductivity is n-type when the compound is in equilibrium with the reference electrode, which may be the parent metal of the compound. It changes via intrinsic to p-type at higher voltages which correspond to lower μ^*_M or higher μ^*_X. This variation is reflected in the slopes in the different regions of the curve.

Note that there exists a close relationship between the σ_e curves of Figs. 7.1b and 7.2b and the curve of Fig. 7.12, of which the *slope* corresponds to the electronic conductivity. Both scales along the abscissas are equivalent because of (7.118). If the total electronic conductivity of Figs. 7.1b and 7.2b were plotted using a linear instead of a logarithmic scale on the ordinate axis, the curve obtained would be the derivative of the curve in Fig. 7.12.

A full curve as indicated by the figure is seldom realized for actual compounds, because of the limitation imposed by the phase diagram. Thus for the cell Ag|AgBr|C, a curve exhibiting n- and p-type branches separated by a plateau [7.56] is found, but for the equivalent cell with CuCl, CuBr or CuI only a p-type branch occurs [7.57]. This observation leads to the conclusion that the copper halides remain p-type conductive (electronically, they are mainly ionic conductors) even in their most reduced form.

To derive quantitative information about the electronic conductivities, the theoretical curve must be matched with the experimental data. σ'_n can be found simply from the height of the current plateau, as is directly evident from Fig. 7.12. The hole conductivity in equilibrium with the reference chemical potential is found from the intercept with the i axis obtained from a $\log i_e$ vs E plot extrapolating the straight line part to $E = 0$. This method is particularly effective if only a p-type branch is exhibited.

An alternate way of analyzing polarization curves, which is useful for cells yielding the complete curve, was given by *Patterson* et al. [7.45]. They plotted $i_e/[1 - \exp(-eE/kT)]$ against $\exp(eE/kT)$ and showed that σ'_n follows from the slope and σ'_p from the intercept with the converted current axis.

It must be remarked that interpretation of the i_e vs E characteristics in terms of electronic conductivity is appropriate only if no material leak from the blocking electrode can occur, e.g., by decomposition and evaporation, or by reaction with the "inert" electrode. Such an effect would lead to an over-estimation of the p-type conductivity. It has been shown that similar experiments with pressed samples of $RbAg_4I_5$ give information about iodine diffusion through the electrolyte towards the silver electrode [7.58].

The information obtainable from the polarized cell measurements described can be related to the electronic bandgap. We note that the product of σ_n and σ_p at any partial pressure, and thus also in the equilibrium situation with the reference electrode, can be expressed in terms of mobilities, state densities and bandgap, according to (7.34) and (7.35), when also (7.6b) is taken into account.

$$\sigma'_n \sigma'_p = e^2 N_c N_v u_n u_p \exp(-eB/kT). \tag{7.124}$$

Thus, by assuming reasonable values for mobilities and effective masses, an estimation of the electronic bandgap is possible.

Alternatively, with a known bandgap mobilities can be estimated.

The method of the static polarized cell has been applied to many electrolytic compounds. Examples are silver halides [7.56, 59], copper halides [7.57], thallium bromide [7.60] and lead chloride [7.61], but the method works also with oxides like ZrO_2 or ThO_2 [7.45, 48, 62].

In principle, the polarized cell technique works also for ternary compounds. However, because of the extra degree of freedom, provision must be made to fix one more chemical potential, or to fix a composition parameter, so that the system becomes quasi-binary. This has sometimes been neglected in results reported in the literature. Data are available for Ag_3SBr and Ag_3SI [7.63–65], and for $AgAl_{11}O_{17}$ [7.66].

7.3.5 The Polarized Cell Technique (Dynamic)

So far we have discussed only methods for the determination of partial electronic *conductivities*. These are proportional to the product of charge carrier concentration and mobility (or diffusion coefficient). We shall now concentrate on

that the complete Fourier series solution of Fick's second law must be used, and secondly, that after differentiation of this solution to obtain the concentration gradient $x=0$ is inserted instead of $x=L$. The concentration profile at a time t after an increase of pressure at $x=L$ following an equilibrium situation with uniform hole concentration p' can then be calculated from [7.39].

$$p = p' + p''x/L + 2p''/\pi \sum_{m=1}^{\infty}$$
$$\cdot [1/m(-1)^m \sin(m\pi x/L) \exp(-D_p m^2 \pi^2 t/L^2)]. \tag{7.131}$$

Here it is assumed that $p'' \gg p'$. From this expression the hole current that is equivalent to the oxygen evolution at $x=0$ follows as

$$|i_p| = eD_p(\partial p/\partial x)_{\text{at } x=0}$$
$$= eD_p p''/L[1 + 2\sum_{m=1}^{\infty}(-1)^m \exp(-D_p m^2 \pi^2 t/L^2)]. \tag{7.132}$$

For large times this flow approaches the stationary value $eD_p p''/L$ corresponding to that given by (7.114) for small σ'_n.

From the concentration profiles it is recognized that it takes some time before the concentration change induced at the right-hand interface has propagated sufficiently to the left to cause gas evolution at the low pressure side. This delay is clearly illustrated in the experimental curve given in Fig. 7.15, which shows the pressure increase with time of a closed volume, separated by a zirconia wall from a gas space where the pressure was increased from 0.1 Torr to 760 Torr at $t=0$. The delay period can be characterized by the time $\tau_{1/2}$ it takes to reach half the stationary flow rate. It is the t-value for which the summation in (7.132) reaches the value $-1/4$. Numerical calculation shows that

$$\tau_{1/2} = 0.1387(L^2/D_p). \tag{7.133}$$

This expression allows determination of D_p from the time delay.

This method has been used by *Smith* et al. [7.77] and by *Heyne* and *Beekmans* [7.39] to determine the chemical diffusion coefficient of stabilized zirconia.

To conclude, a method will be discussed which deserves attention because it can be performed on almost any type of sample. Where the permeation method requires high temperature sealings or, preferably, tube shaped samples, the absorption/desorption technique to be discussed below needs only samples with a well-defined surface area. It is carried out as follows.

A flat, disc shaped, sample of known thickness and surface area (but a spherical sample may also be used) is first equilibrated at a certain partial pressure. Then the reaction vessel is quickly evacuated, the pump closed, and the subsequent gas desorption is followed by recording of the gradually increasing pressure as a function of time. Alternatively, the desorption process (or an equivalent absorption in the reversed experiment) can be followed by means of a thermo-balance [7.78].

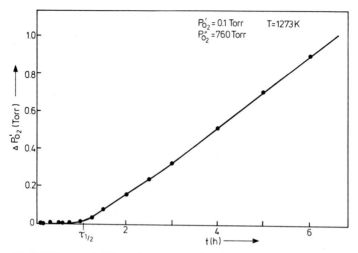

Fig. 7.15. Pressure buildup at low pressure side of oxygen permeation cell after pressure increase to 1 atm at $t=0$. At $\tau_{1/2}$ the slope of the curve (= permeation rate) is one half of the stationary slope [7.39]

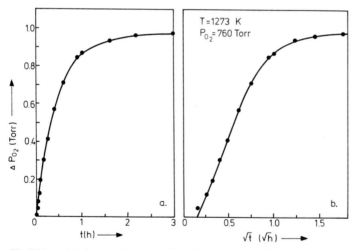

Fig. 7.16a and b. Oxygen desorption from zirconia sample after fast evacuation, (a) linear time scale, (b) square root time scale [7.39]

A solution of the diffusion equation applicable to this case is given by *Barrer* [Ref. 7.70, Eq. 57]. It can be shown that the total quantity of gas desorbed from both sides of the sample after a time t can be described by [7.39]

$$Q_t = 2.26(D_p t)^{1/2} . \tag{7.134}$$

Here Q_t is expressed in terms of the equivalent electrical charge. The equation is an approximation for not too long times. The experimental curve represented in

Fig. 7.16 confirms that a straight line is obtained when Q_t is plotted against $t^{1/2}$. The slope of this line yields the hole diffusion coefficient. From the saturation value of Q_t at $t = \infty$ the hole concentration can be found.

As a final remark, we state that the gas exchange and penetration phenomena discussed have consequences for the performance of solid electrolyte cells used as gas sensors. If solid electrolytes existed with zero electronic conductivity, neither penetration nor transient exchange in excess of that equivalent to double layer charging would occur. Then it would be immaterial whether the electrodes are highly reversible or easily polarized. However, with practical electrolytes penetration occurs, and even higher gas fluxes occur at the interfaces during transients. Apart from a possible influence on the sample to be measured, such fluxes cause overvoltages at electrodes which are not fully reversible. This affects the accuracy of the emf, as has been shown by *Fouletier* et al. [7.55]. Because the fluxes during transient situations following pressure variations may be much larger than in the steady state, the errors induced are larger in the former case [7.10], so that the emf responds slowly to pressure variations if the reversibility of the electrodes is bad.

In this chapter we have given a survey of the basic principles underlying the relation between electronic and ionic phenomena in ionic conductors. Furthermore, the application of these principles in practical methods for the study of these materials was reviewed.

It was decided not to try to give a full account of the present state of knowledge about properties of mixed conductors considered from the material point of view. Therefore, particular materials were discussed only insofar as they could be used to illustrate more general principles.

The author expresses the hope that, acting in this way, he could assist and stimulate those active or interested in this field of materials research.

List of Symbols

a	proportionality constant in mass action law
a_i	activity of particle species i
A	symbol for acceptor type of impurity atom
A	cross-sectional area of sample
B	electronic bandgap
B_0	electronic bandgap at $0\,°C$
c_i	concentration of species i expressed in atomic units/cm^3
D_f	diffusion coefficient as defined in Fick's diffusion laws
D_s	coefficient of self-diffusion
D_i	diffusion coefficient of particle species i
D^*	chemical diffusion coefficient
D_a	ambipolar diffusion coefficient
e	symbol for electron, or, used as an index, electronic

e absolute value of the electronic charge

E electromotive force (emf) of a galvanic cell

E_e electronic energy

f occupancy of electronic energy level

F Fermi level

$F_{1/2}$ Fermi integral

h Planck's constant

H energy depth of trapping level

i_k electrical current density carried by particle species k

j_k particle current density of species k

k Boltzmann's constant

L sample thickness

m electronic mass

m_c effective mass of electron in conduction band

m_v effective mass of hole in valence band

M   symbol for metallic component of compound

M_i M-atom on interstitial lattice position

n electron concentration

n_t concentration of trapped electrons

n_0 intrinsic electronic concentration

N_c effective density of states in conduction band

N_v effective density of states in valence band

N_t concentration of trapping centers

p hole concentration

P_{tot} total pressure

P_X partial pressure of component X

P_0 partial pressure of component X at intrinsic point of pure compound MX

P_0' partial pressure of component X at intrinsic point of doped compound MX

P_n partial pressure of component X at which $\sigma_n = \sigma_{ion}$

P_p partial pressure of component X at which $\sigma_p = \sigma_{ion}$

t_i transport number of charge carrier species i

t_n, t_p transport number of electrons or holes

t_{ion} ionic transport number

t_e electronic transport number: $t_e = t_n + t_p$

t.f.   thermodynamic factor

T absolute temperature

u_i mobility of charge carrier of type i (speed at unity field strength)

V   symbol for vacant lattice position

v_0 intrinsic defect concentration

X   symbol for metalloid component of compound

z valency of metal atoms

y absolute value of valency of metalloid atoms

α trapping factor

δ relative deviation from stoichiometric composition

ε dimensionless energy parameter eE_e/kT

η dimensionless Fermi level eF/kT

ϕ electrostatic potential

μ_i chemical potential of charged particles of type $_i$

$\tilde{\mu}_i$ electrochemical potential of charged particles of type $_i$

μ_M^* chemical potential of neutral component M

μ_i^0 standard chemical potential of charged particles of type $_i$

σ_i partial conductivity of charge carrier type $_i$

σ_n, σ_p partial conductivity of electrons and holes

σ_{ion} partial ionic conductivity

σ_e partial electronic conductivity: $\sigma_e = \sigma_n + \sigma_p$

σ_e^0 partial electronic conductivity at intrinsic point

$\tau_{1/2}$ half value time constant

References

7.1   E. A. Guggenheim: *Thermodynamics*, 5th ed. (North-Holland, Amsterdam 1967)

7.2   F. A. Kröger: *The Chemistry of Imperfect Crystals*, 2nd ed. (North-Holland, Amsterdam 1974)

7.3   F. A. Kröger, H. J. Vink: In *Solid State Physics*, Vol. 3, ed. by F. Seitz, D. Turnbull (Academic Press, New York 1956) p. 307

7.4   F. A. Kröger, H. J. Vink: J. Phys. Chem. Solids **5**, 208 (1958)

7.5   N. B. Hannay: *Semiconductors* (Reinhold, New York 1959)

7.6   W. Shockley: *Electrons and Holes in Semiconductors* (Van Nostrand, New York 1950)

7.7   A. C. Smith, J. F. Janak, R. B. Adler: *Electronic Conduction in Solids* (McGraw Hill, New York 1967)

7.8   O. Madelung: In *Physical Chemistry*, Vol. 10, ed. by W. Jost (Academic Press, New York 1970) Chap. 6

7.9   G. Brouwer: Philips Res. Rept. **9**, 366 (1954)

7.10   L. Heyne, D. den Engelsen: J. Electrochem. Soc. **124**, 727 (1977)

7.11   R. E. Carter, W. L. Roth: In *Electromotive Force Measurements in High Temperature Systems*, ed. by C. B. Alcock (Institution of Mining and Metallurgy, London 1968) p. 125

7.12   J. W. Patterson: J. Electrochem. Soc. **118**, 1033 (1971)

7.13   L. Heyne: In *Fast Ion Transport in Solids*, ed. by W. van Gool (North-Holland, Amsterdam 1973) pp. 123–139

7.14   J. McDougall, E. C. Stoner: Phil. Trans. Roy. Soc. London A**237**, 67 (1938)

7.15   J. S. Blakemore: Elec. Commun. **29**, 131 (1952)

7.16   A. J. Rosenberg: J. Chem. Phys. **33**, 665 (1960)

7.17   C. Wagner: J. Chem. Phys. **21**, 1819 (1953)

7.18   H. Rickert: In *Festkörperprobleme*, ed. by O. Madelung (Vieweg, Braunschweig 1967) p. 85

7.19   N. Valverde: Z. Physik. Chem. NF **70**, 113 (1970); **70**, 128 (1970)

7.20   V. L. Bonch-Bruyevich: *The Electronic Theory of Heavily Doped Semiconductors* (Amer. Elsevier, New York 1966)

7.21   V. I. Fistul': *Heavily Doped Semiconductors* (Plenum Press, New York 1969)

7.22   C. Wagner: In *Progress in Solid State Chemistry*, Vol. 6, ed. by H. Reiss (Pergamon, Oxford 1971)

7.23   C. Wagner: In *Advances in Electrochemistry and Electrochemical Engineering*, Vol. 4, ed. by P. Delahay (Interscience, New York 1966) p. 1

7.24 A. D. Le Claire: In *Physical Chemistry*, Vol. 10, ed. by W. Jost (Academic Press, New York 1970) Chap. 5

7.25 C. Wagner: Z. Physik. Chem. B**11**, 139 (1930)

7.26 C. Wagner: Z. Physik. Chem. B**21**, 25 (1933)

7.27 C. Wagner: Z. Physik. Chem. B**32**, 447 (1936)

7.28 W. van Roosbroeck: Phys. Rev. **91**, 282 (1953)

7.29 B. Hartmann, H. Rickert, W. Schendler: Electrochim. Acta **21**, 319 (1976)

7.30 C. Wagner: In *Atomic Movements* (Am. Soc. for Metals, Cleveland, Ohio 1951) p. 153

7.31 B. C. H. Steele: In *Fast Ion Transport in Solids*, ed. by W. van Gool (North-Holland, Amsterdam 1973) pp. 103–122

7.32 P. E. Childs, L. W. Laub, J. B. Wagner, Jr.: Proc. Brit. Ceram. Soc. **19**, 29 (1971)

7.33 R. H. Campbell, W. J. Kass, M. O'Keefe: In *Mass Transport in Oxides*, ed. by J. B. Wachtmann, A. D. Franklin (Nat. Bur. Stand. Spec. Publ. No. 296, 1968) p. 173

7.34 J. H. Shulman, W. D. Compton: *Color Centers in Solids* (Pergamon Press, Oxford 1963)

7.35 P. Fabry, M. Kleitz: In *Electrode Processes in Solid State Ionics*, ed. by M. Kleitz, J. Dupuy (Reidel, Dordrecht, Holland 1976) p. 331

7.36 R. H. Bube: *Photoconductivity in Solids* (Wiley, New York 1960)

7.37 L. Heyne: Philips Techn. Rev. **25**, 120 (1963); **27**, 47 (1966); **29**, 221 (1968)

7.38 M. A. Lampert, P. Mark: *Current Injection in Solids* (Academic Press, New York 1970)

7.39 L. Heyne, N. M. Beekmans: Proc. Brit. Ceram. Soc. **19**, 229 (1971)

7.40 W. F. Chu, H. Rickert, W. Weppner: In *Fast Ion Transport in Solids*, ed. by W. van Gool (North-Holland, Amsterdam 1973) p. 181

7.41 C. Wagner: In *Proc. 7th Meeting Int. Comm. on Electrochem. Thermodynamics and Kinetics* (Butterworth, London 1957) p. 361, Lindau (1955)

7.42 W. L. Worrell, J. L. Iskoe: In *Fast Ion Transport in Solids*, ed. by W. van Gool (North-Holland, Amsterdam 1973) p. 513

7.43 A. A. Vecher, D. V. Vecher: Russ. J. Phys. Chem. **41**, 685 (1967)

7.44 H. Schmalzried: Z. Physik. Chem. NF **38**, 87 (1962)

7.45 J. W. Patterson, E. C. Bogren, R. A. Rapp: J. Electrochem. Soc. **114**, 752 (1967)

7.46 R. A. Giddings, S. R. Gordon: J. Electrochem. Soc. **121**, 793 (1974)

7.47 W. A. Fischer, D. Jahnke: Z. Physik. Chem. NF **69**, 11 (1970)

7.48 T. H. Etsell, S. N. Flengas: J. Electrochem. Soc. **119**, 1 (1972)

7.49 H. Schmalzried: Z. Elektrochemie **66**, 572 (1962)

7.50 D. Yuan, F. A. Kröger: J. Electrochem. Soc. **116**, 594 (1969)

7.51 W. A. Fischer, D. Jahnke: *Metallurgische Elektrochemie* (Verlag Stahleisen, Düsseldorf und Springer, Berlin-Heidelberg-New York 1975)

7.52 K. S. Goto, W. Pluschkell: In *Physics of Electrolytes*, Vol. 2, ed. by J. Hladik (Academic Press, London, New York 1972) Chap. 13, pp. 540–622

7.53 T. H. Etsell, S. N. Flengas: Chem. Rev. **70**, Nr. 3, 339 (1970)

7.54 R. Hartung, H. H. Möbius: Z. Physik. Chem. Lpz. **243**, 133 (1970)

7.55 J. Fouletier, P. Fabry, M. Kleitz: J. Electrochem. Soc. **123**, 204 (1976)

7.56 B. Illschner: J. Chem. Phys. **28**, 1109 (1958)

7.57 J. B. Wagner, C. Wagner: J. Chem. Phys. **26**, 1597 (1957)

7.58 J. E. Oxley, B. B. Owens: In *Power Sources*, Vol. 3, ed. by D. H. Collins (Oriel press, Newcastle upon Tyne 1971) p. 535

7.59 Y. J. van der Meulen, F. A. Kröger: J. Electrochem. Soc. **117**, 69 (1970)

7.60 A. Morkel, H. Schmalzried: J. Chem. Phys. **36**, 3101 (1962)

7.61 J. B. Wagner, C. Wagner: J. Electrochem. Soc. **104**, 509 (1957)

7.62 L. D. Burke, H. Rickert, R. Steiner: Z. Physik. Chem. NF **74**, 146 (1971)

7.63 B. Reuter, K. Hardel: Ber. Bunsenges. Physik. Chem. **70**, 82 (1966)

7.64 N. Valverde: Z. Physik. Chem. NF **75**, 1 (1971)

7.65 T. Takahashi, O. Yamamoto: Electrochim. Acta **11**, 779 (1966)

7.66 M. S. Whittingham, R. A. Huggins: J. Electrochem. Soc. **118**, 1 (1971)

7.67 J. B. Wagner, Jr.: In *Fast Ion Transport in Solids*, ed. by W. van Gool (North-Holland, Amsterdam 1973) p. 489

7.68 A.V.Joshi, J.B.Wagner,Jr.: J. Phys. Chem. Solids **33**, 205 (1972)

7.69 K.Weiss: Z. Physik. Chem. NF **59**, 242 (1968)

7.70 R.M.Barrer: *Diffusion in and through Solids* (Cambridge University Press, New York 1951)

7.71 H.S.Carlslaw, J.C.Jaeger: *Conduction of Heat in Solids* (Clarendon Press, Oxford 1959)

7.72 J.Crank: *Mathematics of Diffusion* (Clarendon Press, Oxford 1956)

7.73 A.V.Joshi: Thesis North-Western University (1972)

7.74 J.B.Wagner: In *Electrode Processes in Solid State Ionics*, ed. by M.Kleitz, J.Dupuy (Reidel, Dordrecht, Holland 1976) pp. 185–218

7.75 D.O.Raleigh: Z. Physik. Chem. NF **63**, 319 (1969)

7.76 A.V.Joshi: In *Fast Ion Transport in Solids*, ed. by W. van Gool (North-Holland, Amsterdam 1973) p. 173

7.77 A.W.Smith, F.W.Meszaros, C.D.Amata: J. Am. Ceram. Soc. **49**, 240 (1966)

7.78 J.M.Wimmer, N.M.Tallan: In *Conduction in Low-Mobility Materials*, ed. by N.Klein, D.S.Tannhauser, M.Pollak (Taylor and Francis Ltd. London 1971) p. 41

Additional References with Titles

Chapter 3

U.v.Alpen, J.Fenner, J.D.Marcoll, A.Rabenau: Structural aspects and high partial Cu^+-Ionic conductivity in compounds of CuTeX (X = Cl, Br, I). Electrochim. Acta **22**, 801 (1977)

R.L.Ammlung, D.F.Shriver, M.Kamimoto, D.H.Whitmore: Conductivity and Raman spectroscopy of new indium (I) and thallium (I) ionic conductors, In_4CdI_6, In_2ZnI_4, Tl_2ZnI_4 and the related compound Tl_4CdI_6. J. Solid State Chem. **21**, 185 (1977)

V.S.Borokov, A.K.Ivanov-Shitz: Conductivity and phase transitions in a solid electrolyte $RbAg_4I_5$. Electrochim. Acta **22**, 713 (1977)

W.Bührer, W.Hälg: Crystal structure of high-temperature cuprous iodide and cuprous bromide. Electrochim. Acta **22**, 701 (1977)

M.L.Knotek, C.H.Seager: The absence of a measurable Hall effect in the superionic conductor $RbAg_4I_5$. Solid State Commun. **21**, 625 (1977)

J.-M.Reau, C.Lucat, G.Campet, J.Claverie, J.Portier: New anionic conducting fluorides. Electrochim. Acta **22**, 761 (1977)

M.B.Salamon: Jahn-Teller-like model for the 208-K phase transition in the solid electrolyte $RbAg_4I_5$. Phys. Rev. B**15**, 2236 (1977)

K.Shahi: Transport studies on superionic conductors. Phys. Stat. Sol. (a) **41**, 11 (1977)

T.Takahashi, N.Wakabayashi, O.Yamamoto: Solid state ionics: the electrical conductivity in the $Ag_{1-x}Cu_xI$-substituted ammonium iodide double salts. J. Solid State Chem. **21**, 73 (1977)

Chapter 5

R.D.Armstrong, R.A.Burnham, P.M.Willis: The breakdown of β-alumina at positive and negative potentials. J. Electroanal. Chem. **67**, 111 (1976)

A.S.Barker, Jr., J.A.Ditzenberger, J.P.Remeika: Lattice vibriations and ion transport spectra in β-alumina. I. Infrared spectra. Phys. Rev. B, **14**, 386 (1976); — II. Microwave spectra. Phys. Rev. B **14**, 4254 (1976)

E.Bergmann, H.Tannenberger, M.Voinov: A theoretical model for the exchange of sodium ions between propylene carbonate and β-Al_2O_3. Electrochim. Acta **22**, 459 (1977)

G.V.Chandrashekhar, L.M.Foster: Ionic conductivity of monocrystals of the gallium analogs of β- and β''-alumina. J. Electrochem. Soc.**124**, 329 (1977)

G.C.Farrington: Na^+ transport across the beta alumina-propylene corbonate interface. J. Electroanal. Chem. **76**, 165 (1977)

L.M.Foster, J.E.Scardefield: Growth of monocrystals of the gallium analog of β''-Al_2O_3 by Na_2O evaporation. J. Electrochem. Soc.**124**, 434 (1977)

R.J.Grant, M.D.Ingram, A.R.West: Anisotropic conductivity effects in β-alumina. J. Electroanal. Chem. **72**, 397 (1976)

M.Green, K.S.Kang: Solid state electrochromic cells: the M-β-alumina/WO_3 system. Thin Solid Films **40**, L19 (1977)

P.E.D.Morgan: Low temperature synthetic studies of beta-aluminas. Mat. Res. Bull. **11**, 233 (1976)

W.L.Roth, R.C.DeVries: Comparison of the thermal stability of β- and β''-alumina at high pressures. J. Solid. State Chem. **20**, 111 (1977)

U.Strom, P.C.Taylor, S.G.Bishop, T.L.Reinecke, K.L.Ngai: Microwave and infrared conductivity of Na-β-alumina; evidence for collective ionic effects. Phys. Rev. B **13**, 3329 (1976)

Formula Index

Subject Index

Applied Physics

A monthly journal

Board of Editors	**S. Amelinckx,** Mol. · **V. P. Chebotayev,** Novosibirsk **R. Gomer,** Chicago, Ill. · **H. Ibach,** Jülich **V. S. Letokhov,** Moskau · **H. K. V. Lotsch,** Heidelberg **H. J. Queisser,** Stuttgart · **F. P. Schäfer,** Göttingen **A. Seeger,** Stuttgart · **K. Shimoda,** Tokyo **T. Tamir,** Brooklyn, N.Y. · **W. T. Welford,** London **H. P. J. Wijn,** Eindhoven
Coverage	application-oriented experimental and theoretical physics:

Solid-State Physics   *Quantum Electronics*
Surface Physics   *Laser Spectroscopy*
Chemisorption   *Photophysical Chemistry*
Microwave Acoustics   *Optical Physics*
Electrophysics   *Integrated Optics*

Special Features	**rapid** publication (3–4 months) **no** page charge for **concise** reports prepublication of titles and abstracts **microfiche** edition available as well
Languages	Mostly English
Articles	original reports, and short communications review and/or tutorial papers
Manuscripts	to Springer-Verlag (Attn. H. Lotsch), P.O. Box 105 280 D-69 Heidelberg 1, F.R. Germany

Place North-American orders with:
Springer-Verlag New York Inc., 175 Fifth Avenue, New York. N.Y. 10010, USA

Springer-Verlag
Berlin Heidelberg New York

Principles of Magnetic Resonance

2nd, revised and expanded edition.
112 figures. Approx. 360 pages. 1978
(Springer Series in Solid-State
Sciences, Volume 1)
ISBN 3-540-08476-2

Contents: Elements of Resonance. —
Basic Theory. — Magnetic Dipolar
Broadening of Rigid Lattices. —
Magnetic Interactions of Nuclei with
Electrons. — Spin-Lattice Relaxation and
Motional Narrowing of Resonance Lines.
— Spin Temperature in Magnetism and
in Magnetic Resonance. — Double
Resonance. — Advanced Concepts in
Pulsed Magnetic Resonance. — Electric
Quadrupole Effects. — Electron Spin
Resonance. — Summary. — Appendices.

Dynamics of Solids and Liquids by Neutron Scattering

Editors: S.W.Lovesey, T.Springer
156 figures, 15 tables. XI, 379 pages. 1977.
(Topics in Current Physics, Volume 3)
ISBN 3-540-08156-9

Contents:
S.W.Lovesey, Introduction. — *H.G.Smith*,
N.Wakabayashi, Phonons. — *B.Dorner*,
R.Comès, Phonons and Structural Phase
Transformations. — *J.W.White*, Dyna-
mics of Molecular Crystals, Polymers,
and Adsorbed Species. — *T.Springer*,
Molecular Rotations and Diffusion in
Solids, in Particular Hydrogen in Metals.
— *R.D.Mountain*, Collective Modes in
Classical Monoatomic Liquids. —
S.W.Lovesey, J.M.Loveluck, Magnetic
Scattering.

Solid-State Physics

75 figures. III, 153 pages. 1974.
(Springer Tracts in Modern Physics,
Volume 74)
ISBN 3-540-06946-1

Contents: *G.Bauer*, Determination of
Electron Temperatures and of Hot Electron Distribution Functions in Semiconductors.
G.Borstel, *H.J.Falge*, *A.Otto*, Surface
and Bulk Phonon-Polaritons Observed
by Attenuated Total Reflection.

Neutron Physics

40 figures, 11 tables. VII, 135 pages. 1977.
(Springer Tracts in Modern Physics,
Volume 80)
ISBN 3-540-08022-8

Contents: *L.Koester*, Neutron Scattering
Lengths and Fundamental Neutron
Interactions. *A.Steyerl*, Very Low Energy
Neutrons.

Springer-Verlag
Berlin Heidelberg
New York